Atmospheric Chemistry and Physics

Atmospheric Chemistry and Physics

Editor

Sanjay Deshmukh

Atmospheric Chemistry and Physics

Edited by **Sanjay Deshmukh**

Printed in 2017

ISBN: 978-1-68117-133-3

Library of Congress Control Number: 2015936554

© 2016 by
SCITUS Academics LLC,
616, Corporate Way, Suite 2, 4766,
Valley Cottage, NY 10989

www.scitusacademics.com

Contents

Preface

Atmospheric chemistry is a branch of atmospheric science in which the chemistry of the Earth's atmosphere and that of other planets is studied. It is a multidisciplinary field of research and draws on environmental chemistry, physics, meteorology, computer modeling, oceanography, geology and volcanology and other disciplines. Research is increasingly connected with other areas of study such as climatology. The composition and chemistry of the atmosphere is of importance for several reasons, but primarily because of the interactions between the atmosphere and living organisms. The composition of the Earth's atmosphere changes as result of natural processes such as volcano emissions, lightning and bombardment by solar particles from corona. It has also been changed by human activity and some of these changes are harmful to human health, crops and ecosystems.

Editor

Multi-Stage Open Peer Review: Scientific Evaluation Integrating the Strengths of Traditional Peer Review with the Virtues of Transparency and Self-Regulation

Ulrich Pöschl

Max Planck Institute for Chemistry, Mainz, Germany

ABSTRACT

The traditional forms of scientific publishing and peer review do not live up to all demands of efficient communication and quality assurance in today's highly diverse and rapidly evolving world of science. They need to be advanced and complemented by interactive and transparent forms of review, publication, and discussion that are open to the scientific community and to the public. The advantages of open access, public peer review, and interactive discussion can be efficiently and flexibly combined with the strengths of traditional scientific peer review. Since 2001 the benefits and viability of this approach are clearly demonstrated by the highly successful interactive open access journal Atmospheric Chemistry and Physics (ACP, www.atmos-chem-phys.net) and a growing number of sister journals launched and operated by the European Geosciences Union (EGU,www.egu.eu) and the open access publisher Copernicus (www.copernicus.org). The interactive open access journals are practicing an integrative multi-stage process of publication and peer review combined with interactive public discussion, which effectively resolves the dilemma between rapid scientific exchange and thorough quality assurance. Key features and achievements of this approach are: top quality and impact, efficient self-regulation and low rejection rates, high attractivity and rapid growth, low costs, and financial sustainability. In fact, ACP and the EGU interactive open access sister journals are by most if not all standards more successful than comparable scientific journals with traditional or alternative forms of peer review (editorial statistics, publication statistics, citation statistics, economic costs, and sustainability). The high efficiency and predictive validity of multi-stage open peer review have been confirmed in a series of dedicated studies by evaluation experts from the social sciences, and the same or similar concepts have recently also been adopted in other disciplines, including the life sciences and economics. Multi-stage open peer review can be flexibly adjusted to the needs and peculiarities of different scientific communities. Due to the flexibility and compatibility with traditional structures of scientific publishing and peer review, the multi-stage open peer review concept enables efficient evolution in scientific communication and quality assurance. It has the potential for swift replacement of hidden peer review as the standard of scientific quality assurance, and it provides a basis for open evaluation in science.

INTRODUCTION

The traditional ways of scientific publishing and peer review do not live up to the needs of efficient communication and quality assurance in today's highly diverse and rapidly developing world of science. Besides high profile cases of scientific fraud, science, and society are facing a flood of carelessly prepared scientific papers that are locked away behind subscription barriers, dilute rather than enhance scientific knowledge, lead to a waste of resources and impede scientific and societal progress. On the other hand, the spread of innovative ideas and concepts is often delayed by inertia and obstruction in the hidden review process of traditional mainstream scientific journals (Pöschl, 2004).

Open access to scientific research publications is desirable for many educational, economic, and scientific reasons (Max Planck Society, 2003; David and Uhlir, 2005; European Commission and German Commission for UNESCO, 2008), and it provides major opportunities for the improvement of scientific communication, quality assurance, and evaluation (Bodenschatz and Pöschl, 2008;Pöschl and Koop, 2008; Pöschl, 2010b):

- Open access is fully compatible with traditional peer review, and in addition it enables interactive and transparent forms of review and discussion open to all interested members of the scientific community and the public (open peer review).

- Open access gives reviewers more information to work with, i.e., it provides unlimited access to relevant publications across different scientific disciplines and communities (interdisciplinary scientific discussion and quality assurance).

- Open access facilitates the development and implementation of new metrics for the impact and quality of scientific publications (combination of citation, download/usage, commenting, and ranking by various groups of readers and users, respectively; Bollen et al., 2009).

- Open access helps to overcome the obsolete monopoly/oligopoly structures of scientific publishing and statistical analysis of publication contents and citations/references, which are limiting the opportunities for innovation in scientific publishing and evaluation.

As demonstrated below, the effects and advantages of open access, public review, and interactive discussion can be efficiently and flexibly combined with the strengths of traditional scientific publishing and peer review (Pöschl, 2009a, 2010a,b). Unlike other, more radical proposals of how to change and improve scientific quality assurance, the interactive open access publishing approach introduced by the international scientific journal Atmospheric Chemistry and Physics (ACP) conserves the strengths of traditional peer review while overcoming its major weaknesses. This approach is compatible with the structures of traditional scientific publishing and quality assurance, and thus it enables an efficient transition from the operational but sub-optimal past of subscription-based journals and hidden peer review to the future of free exchange and transparent evaluation of scientific information on the internet.

MULTI-STAGE OPEN PEER REVIEW

So far, the arguably most successful alternative to the closed peer review of traditional scientific journals is the multi-stage open peer review practiced by ACP and a growing number of interactive open access sister journals of the European Geosciences Union (EGU) and Copernicus Publications (Pöschl, 2010b). As detailed below (see Atmospheric Chemistry and Physics and the European Geosciences Union), ACP is by most if not all standards more successful than comparable scientific journals with traditional or alternative forms of peer review (editorial statistics, publication statistics, citation statistics, economic costs, and sustainability). The multi-stage open peer review of ACP is based on a two-stage process of open access publishing combined with multiple steps of peer review and interactive public discussion as illustrated in Figure 1.

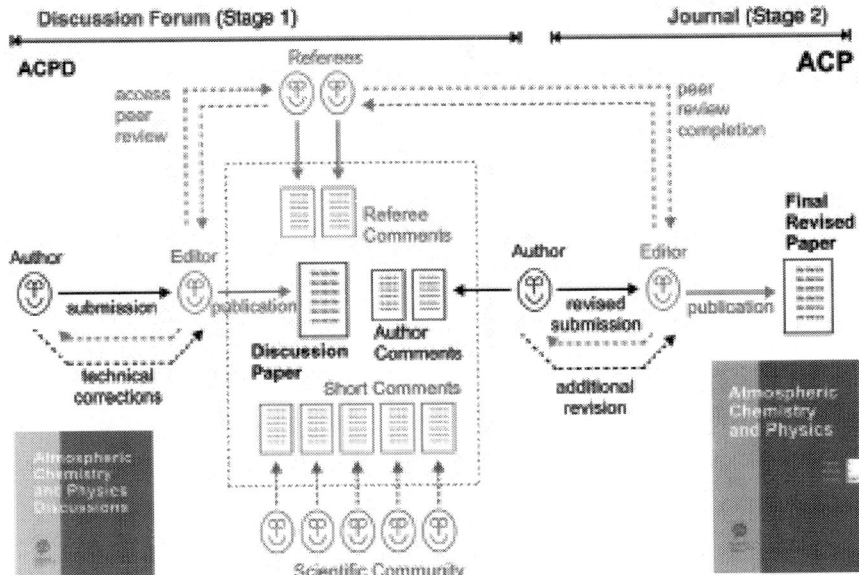

Figure 1: Multi-stage open peer review as practiced in the scientific journal Atmospheric Chemistry and Physics (ACP) and its discussion forum Atmospheric Chemistry and Physics Discussions (ACPD). Solid and dashed arrows indicate required and optional processes and interactions between author, editor, referees, and scientific community.

In the first stage, manuscripts that pass a rapid pre-screening (access review) are immediately published as "discussion papers" in the journal's discussion forum (Atmospheric Chemistry and Physics Discussions, ACPD). They are then subject to interactive public discussion for a period of 8 weeks, during which the comments of designated referees, additional comments by other interested members of the scientific community, and the authors' replies are also published alongside the discussion paper. While referees can choose to sign their comments or remain anonymous, comments by other scientists (registered readers) are automatically signed. In the second stage, manuscript revision and peer review are completed in the same way as in traditional journals (with further rounds of review and revision where required) and, if accepted, final papers are published in the main journal. To provide a lasting record of review and to secure the authors' publication precedence, every discussion paper, and interactive comment remains permanently archived and individually citable.

The multi-stage peer review and publication process of ACP effectively resolves the dilemma between rapid scientific exchange and thorough quality assurance, and it offers a win-win situation for all involved parties (authors, referees, editors, publishers, readers/scientific community). The primary positive effects and advantages compared to the traditional forms of publication with closed peer review are:

- The discussion papers offer free speech and rapid dissemination of novel results and original opinions, without revisions that might delay or dilute innovation (authors' and readers' advantage).

- The interactive peer review and public discussion offer direct feedback and public recognition for high quality papers (authors' advantage); they prevent or minimize the opportunity for hidden obstruction and plagiarism (authors' advantage); they provide complete and citable documentation of critical comments, controversial arguments, scientific flaws, and complementary information (referees' and readers' advantage); they reveal deficiencies and deter submissions of carelessly prepared manuscripts, thus helping to avoid/minimize the waste of time and effort for deficient submissions (referees', editors', publishers', and readers' advantage).

- The final revised papers offer a maximum of scientific information density and quality assurance achieved by full peer review (with optional anonymity of referees) and revisions based on the referees' comments plus additional comments from other interested scientists (readers' advantage).

Readers who are primarily interested in the quintessence of manuscripts that have been fully peer reviewed and approved by referees and editors can simply focus on the final revised paper (or, indeed, its abstract) published in the journal and neglect the preceding discussion papers and interactive comments published in the discussion forum. Thus the two-stage publication process does not inflate the amount of time required to maintain an overview of final revised papers. On the other hand, readers who want to see original scientific manuscripts and messages before they are influenced by peer review and revision, and who want to follow the scientific discussion between authors, referees, and other interested scientists, can browse the papers and interactive comments in the discussion forum.

The possibility of comparing a final revised paper with the preceding discussion paper and following the interactive peer review and public discussion also facilitates the evaluation of individual publications for non-specialist readers and evaluators. The style and quality of interactive commenting and argumentation provide insights that go beyond, and complement, the information contained in the research article itself.

The multi-stage process of review and publication stimulates scientists to prove their competence via individual high quality papers and their discussion, rather than just by pushing as many papers as possible through journals with closed peer review and no direct public feedback and recognition for their work. Authors have a much stronger incentive to maximize the quality of their manuscripts prior to submission for peer review and publication, since experimental weaknesses, erroneous interpretations, and relevant but unreferenced earlier studies are more likely to be detected and pointed out in the course of interactive peer review and discussion open to the public and all colleagues with related research interests.

Moreover, the transparent review process prevents authors from abusing the peer review process by delegating some of their own tasks and responsibilities to the referees during review and revision behind the scenes. Referees often make substantial contributions to the quality of scientific papers, but in traditional closed peer review their input rarely receives public recognition. The full credit for the quality of a paper published in a traditional journal generally goes to the authors, even when they have submitted a carelessly prepared manuscript that has taken a lot of time and effort on the part of the referees, editors, and publishers to turn it into a good one. While peer review depends crucially on the availability and performance of referees, it has traditionally offered little reward for those providing careful and constructive reviews. In public review, however, referees' arguments are publicly heard and, if comments are openly signed, referees can also claim authorship for their contribution.

Note that most of the effects and advantages outlined above are not fully captured by alternative approaches where interactive commenting and public discussion occur only after formal peer review and final publication of scientific papers or where the discussion paper and interactive comments are removed after publication of the final revised paper (see Key features of multi-stage open peer review as practiced by ACP).

Overall, the interactive open access publishing philosophy emphasizes the value of free speech and efficient public exchange and scrutiny of scientific results in line with the principles of critical rationalism and open societies. Accordingly, editors and referees are supposed to critically comment and evaluate manuscripts, to help authors improve their manuscripts, and to eliminate clearly deficient manuscripts. However, authors shall not be forced to adopt the editors' or referees' views and preferences. Instead, the readers shall be able to make up their own mind in view of the public review and discussion. In case of doubt, editorial decisions shall favor free speech of scientists, and in the end, scientific progress; history shall tell if – or to which degree – they were right. In scientific research, the line between fundamental flaws and major innovations can be fine, and the multi-stage process of interactive open access publishing and peer review enables efficient balancing and differentiation between potentially misleading hypotheses and innovative theories even in highly controversial cases (Pöschl, 2004, 2010b).

ATMOSPHERIC CHEMISTRY AND PHYSICS AND THE EUROPEAN GEOSCIENCES UNION

The interactive open access journal Atmospheric Chemistry and Physics (ACP[1]), founded in 2001, demonstrates that multi-stage open peer review enables much more efficient quality assurance than traditional closed peer review. ACP is run by the European Geosciences Union (EGU[2]), the open access publisher Copernicus[3], and a globally distributed network of scientists (~130 co-editors coordinated by an executive committee of five). Manuscripts are normally handled by an editor who is familiar with the specific subject area of the submitted work and independently guides the review process. Details about the largely automated handling and editor assignment of submitted manuscripts are given below (see Key features of multi-stage open peer review as practiced by ACP) and on the journal website. The origin and development of interactive open access publishing as practiced by ACP and EGU/Copernicus are specified in a recent anniversary publication (Pöschl, 2010c, 2011; Copernicus, 2011)[4].

Currently ACP publishes about 800 papers per year (~13,000 double column print pages), which is similar to the volume of traditional major journals in the fields of chemistry and physics (ISI Science Citation Index, Journal Citation Report, 2010). On average, each paper receives four interactive comments, and about one in five papers receives a comment from the scientific community in addition to the comments from designated referees. In total, there are typically 0.5 pages of interactive comments per page of original discussion paper, i.e., the volume of interactive comments amount to as much as ~50% of the volume of discussion papers. The interactive comments show the full spectrum of opinions in the scientific community, ranging from harsh criticism to open applause (sometimes for the same discussion paper), and they provide a wealth of additional information and evaluation that is available to everyone.

About three out of four referee comments are posted without the referee's name, showing that most referees in the scientific community of ACP prefer anonymity. There are, however, interesting differences between sub-disciplines: on average about 20% of theoreticians and computer modelers sign their referee comments, while only 10% of the laboratory and field experimentalists do so. It appears that modelers more often provide suggestions and ideas for which they like to claim authorship as a reward. The anonymous referee comments are generally also very constructive and substantial. The ACP editors do not actively moderate the public discussions but reserve the right to delete abusive or inappropriately worded comments. Out of the nearly 20,000 interactive comments that have been posted so far, only a handful were removed or replaced because of inappropriate wording, which demonstrates efficient self-regulation by transparency.

Some colleagues have expressed concerns that referees may lose their independence by having access to the comments from fellow referees and from the public. Indeed, referees with limited capacities occasionally seem to duplicate or refer to earlier comments without making up their own mind, but this is fairly easy to recognize and to take into account by editors and readers. Much more often, however, referees constructively build on or contradict earlier comments, which enhances the efficiency of review and discussion substantially.

In theory, the independence of referees could be maintained by keeping submitted referee comments non-public until all referees have submitted their comments and these are all together published at the same time. In practice, however, this would cause unnecessary delays ("waiting for the last referee") and stifle rather than promote interactive discussion. Overall, experience shows that the advantages of enabling direct interaction between referees clearly outweigh the disadvantages.

The average rate of public commenting in addition to the designated referees' and authors' comments specified above (~20%) may appear low at first sight. It is, however, by an order of magnitude (factor ~10) higher than in journals with post-peer review online commenting and in traditional journals without online commenting (about 1–2%; Müller, 2008; Pöschl and Koop, 2008;Pöschl, 2010b). Discussion papers reporting controversial findings or innovations attract many interactive comments (up to 30 and more, see "Most commented papers" in the ACPD online library[5]. As expected, non-controversial papers usually elicit comments only from the designated referees. Why would scientists invest effort and time commenting on papers which they find interesting but not controversial?

In most scientific disciplines and journals (certainly in the fields of physics, chemistry, and biology with which the author is well acquainted) it is notoriously difficult to assign a couple of competent referees to every manuscript submitted for publication. In fact, this is the main bottleneck of peer review and scientific quality assurance, and most journal editors have to apply lots of manpower and electronic tools (invitation and reminder emails, etc.) to obtain a couple of referee comments per manuscript. Accordingly, the initiators and editors of ACP are quite satisfied with the overall number and volume of interactive comments. Higher rates of commenting were not expected and are not required to stimulate self-regulation mechanisms of scientific quality assurance (Pöschl, 2004,2010a,b).

The editorial and citation statistics of ACP clearly demonstrate that multi-stage open peer review indeed facilitates and enhances scientific communication and quality assurance. The journal has relatively low rejection rates (~15% as opposed to ~50% in comparable traditional journals, Schultz, 2010), but only a few years after its launch ACP had already achieved top reputation and visibility in the scientific community. Accordingly, it quickly reached and maintained one of the highest ISI impact factors of several 100 journals indexed across the disciplines of atmospheric sciences, geosciences, and environmental sciences (JIF ≈ 5). These figures clearly confirm that anticipation of public peer review and discussion deters authors from submitting low-quality manuscripts and, thus, relieves editors and referees from spending too much time on deficient submissions. This is particularly important, because refereeing capacities are the most limited resource in scientific publishing and quality assurance. The high efficiency, robustness, and predictive validity of the multi-stage open peer review process of ACP have been confirmed in a series of dedicated studies by evaluation experts from the social sciences (Bornmann and Daniel, 2010a,b; Bornmann et al., 2010,2011a,b).

Since its launch in 2001, the number of articles published in ACP has increased rapidly. The high and increasing rates of submission, publication, and citation show that the scientific community values the open access, high quality, and interactive discussions of ACP. They confirm that there is a demand for improved scientific publishing and quality assurance, and that the interactive open access journal concept of ACP meets this demand. Today ACP is the largest journal in the field of atmospheric sciences and one of the largest across the fields of environmental and geosciences, offering at the same time top visibility and low rejection rates (2/5 year impact factors 5.4/5.8, rejection rate 15%, 12,000 pages in 2010). The combination of top visibility with high volume and low rejection rate, i.e., high efficiency by self-regulation, is a fairly unique achievement in the world of scientific publishing, where the most visible journals traditionally had relatively small volumes and high rejection rates (Copernicus, 2011; Pöschl, 2011).

Following up on the successful development of ACP, the EGU, and Copernicus have launched and are operating over a dozen of interactive open access sister journals in the geosciences and related disciplines, and more are in the pipeline[6]:

- Atmospheric Chemistry and Physics (ACP)[7],
- Atmospheric Measurement Techniques (AMT)[8],
- Biogeosciences (BG)[9],
- Climate of the Past (CP)[10],
- Drinking Water Engineering and Science (DWES)[11],

- Earth System Dynamics (ESD)[12],
- Earth System Science Data (ESSD)[13],
- Geoscientific Instrumentation, Methods and Data Systems (GI, geoscientific-instrumentation-methods-and-data-systems.net),
- Geoscientific Model Development (GMD)[14],
- Hydrology and Earth System Sciences (HESS)[15],
- Ocean Science (OS)[16],
- Social Geography (SG)[17],
- Solid Earth (SE)[18],
- The Cryosphere (TC)[19].

Figure 2 illustrates the growth of ACP and the other EGU interactive open access journals over the past decade[20]. The wide range of different topics and scientific communities covered by the EGU interactive open access journals demonstrates that multi-stage open peer review is suitable for any kind of topical scientific journal. For example, the community of cryospheric sciences is much smaller than that of atmospheric sciences, but the development of the cryospheric science journal (TC) proceeds at least as well as that of the atmospheric science journals (ACP and AMT). The first journal impact factor of TC was already the highest in its field. The journal Hydrology and Earth System Sciences (HESS) had already existed as a subscription-based journal with traditional peer review before it was converted into an interactive open access journal. Soon after the transition, the journal experienced a substantial increase of submissions, publications, and citations, demonstrating that traditional journals can be successfully converted into interactive open access journals. Three other open access journals published by EGU (Annales Geophysicae, Natural Hazards, and Earth System Sciences, Non-linear Processes in Geophysics) have maintained traditional peer review up to now. In view of the more successful development of the interactive open access journals, however, they are planning to introduce multi-stage open peer review as well.

FIGURE 2

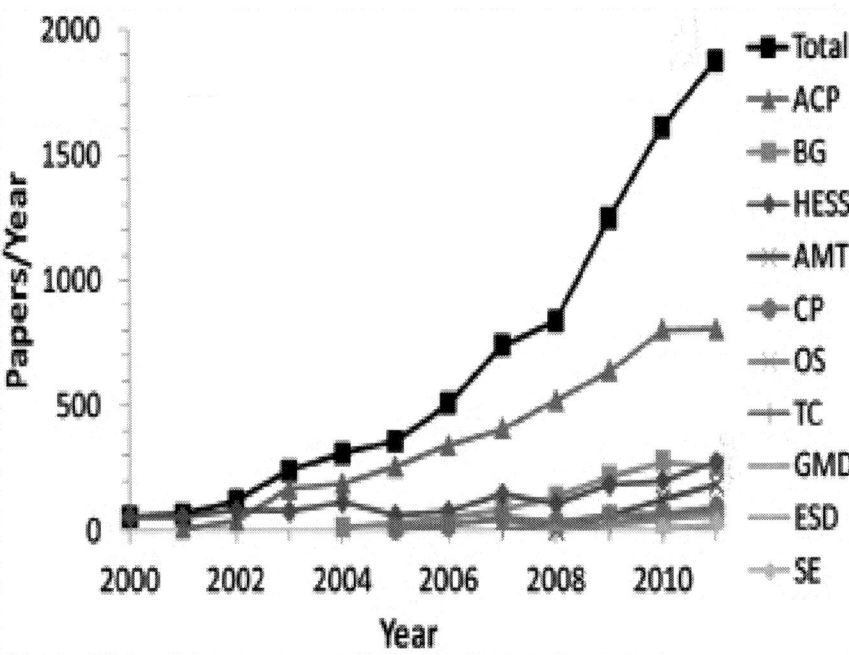

Figure 2: Number of papers published per year in the interactive open access journals of the European Geosciences Union (EGU).

The multi-stage open peer review concept of ACP has also been adopted by the e-journal Economics[21] which was launched in 2007 and involves some of the most prominent institutions and scientists in the field of economics. Alternative concepts of public peer review and interactive discussion are pursued by the open access publications Journal of Advances in Earth System Modeling (JAMES; since 2008)[22], PLoS One[23], Biology Direct[24], Electronic Transactions of Artificial Intelligence (ETAI; since 1997)[25], and Journal of Interactive Media in Education (JIME; since 1996)[26]. Differences between the peer review concepts of these publications and ACP will be addressed and discussed below (see Key features of multi-stage open peer review as practiced by ACP and Comparison to Earlier Initiatives with Two- or Multi-Stage Open Peer Review). In short, approaches where interactive commenting and public discussion are not fully integrated with formal peer review by designated referees tend to be less successful.

FINANCING AND SUSTAINABILITY OF INTERACTIVE OPEN ACCESS PUBLISHING

Atmospheric Chemistry and Physics and its EGU/Copernicus sister journals prove not only the scientific but also the economic viability and sustainability of interactive open access publishing and peer review. The journals were launched and are operated by the independent scientific society EGU and by the small commercial enterprise Copernicus without public subsidies, private donations, or venture capital as involved in the start-up and operation of other successful open access publishers like PLoS and BioMed Central. After several years of operation, ACP and its sister journals have recovered the financial investments of EGU and Copernicus during the start-up phase, and they now deliver a surplus which supports the start-up of new journals by the scientific society as well as a healthy growth of the commercial publisher generating dozens of new jobs.

By developing and applying efficient software tools for the handling of manuscripts (submission, peer review and commenting, typesetting/ production, and distribution), and because minimal time and effort is wasted on carelessly prepared papers (high quality of submissions and low rejection rates as detailed above), Copernicus is able to produce top quality publications at comparatively low cost. The publication service charges are of the order of one hundred Euros per page in final double column format, i.e., about one thousand Euros for an average paper with a length of about ten pages. The service charges cover the review support from the editorial office, free use of color figures and online supplementary materials (data, pictures, movies etc.), typesetting of both the discussion and the final version of the paper, archiving and distribution of papers, and interactive comments (maintenance of websites and servers, electronic copies for open archives, paper copies for copyright libraries, etc.) and overheads. In agreement between the publisher (Copernicus) and the scientific society (EGU council and publications committee), the service charges are adjusted to cover the full costs of publishing, including all the tasks and services outlined above, and to generate a modest surplus for the scientific union: ~10% of the annual financial turnover (currently about three million Euros). The surplus is re-invested in publication development (new journals and services) and it helps to run the membership and outreach activities of EGU, which is a non-profit organization. Like the other scientific officers of the union, editors do their work unpaid on a purely voluntary basis. Following up on the questions and suggestions of a reviewer of this manuscript, I would like to clarify that neither I nor any other editor of ACP and the other EGU interactive open access journals have had any income from the journals that we edit as a voluntary community service. In fact, we pay regular registration fees of up to 500 EUR to attend the annual general assembly and scientific conference of our union (EGU), where the editorial board meetings take place. The separation of financial and scientific interests seems important in the context of peer review, and the ACP/EGU experience demonstrates that a purely voluntary approach on the scientific editors' side is sustainable and compatible with efficient operation of open access journals by a commercial publisher.

For each paper published in ACP, the service charges are levied from the authors or paid by their scientific institution. Since 2008 the German Max Planck Society (MPG)[27] and the French Centre National de Recherche Scientifique (CNRS)[28] have contracts with Copernicus for automated coverage of service charges incurred by their scientists. Other scientific institutions are likely to follow these examples, and many national and international research organizations and funding agencies pursue complementary ways of covering open access service charges for their scientists and projects. Like other open access publishers, Copernicus, and EGU are ready to cover the costs for up to 10% of the papers published each year, if the authors are unable to pay the service charges (e.g., authors without institutional support or institutions from less developed countries). Currently, most papers published in ACP originate from Europe (~50%) and North America (~30%), but the proportion of papers originating from Asia and other regions is increasing.

The ACP open access publication service charges compare quite favorably with the charges levied by other comparable scientific journals and publications:

- Other major open access publishers such as BioMed Central and the Public Library of Science (PLoS) typically charge more than 1,000 EUR for traditional single-stage journal publications.

- Traditional publishing groups like Springer charge 2,000 EUR for making individual publications in traditional subscription journals freely available online ("open choice"), i.e., they levy 2,000 EUR per online open access paper in addition to charging libraries and other subscribers for access to the journal in which it appears.

- In the traditional scientific publishing business, where some journals do not only limit access to subscribers or sell articles on a pay-per-view basis but also request additional publication charges from authors (up to several hundred US dollars per page or color figure), the total turnover, and public costs amount to several thousand US dollars per paper. The annual turnover of journal publishing in the sector of science, technology, and medicine (STM) amounts to around seven billion USD per year, and some of the traditional publishers – led by Elsevier with a market share of about 30% – make operating profits of up to 30% and more. Note that a large proportion of the turnover and profit in STM publishing comes from packaging and selling publicly funded research results that are peer reviewed by publicly funded scientists to publicly funded institutions of education and research (Economist Academic Publishing, 2011; Golden and Schultz, 2012).

In view of these facts, ACP authors and the ACP scientific community have had little difficulty in accepting and paying average service charges of about one thousand Euros per paper to make ACP and its sister journals sustainable. Overall, ACP and its interactive open access sister journals prove that top quality (interactive) open access publishing and peer review can be realized and sustained by scientific societies and (small) commercial publishers with tightly limited budgets and without public subsidies, private donations or venture capital. Indeed, ACP, EGU, and Copernicus demonstrate how STM publishing at large can and will hopefully soon manage a swift transition from the past of print-based subscription barriers into the future of an internet-based open access environment.

KEY FEATURES OF MULTI-STAGE OPEN PEER REVIEW AS PRACTICED BY ACP

The following key features of the ACP multi-stage open peer review system help ensure maximum efficiency of scientific exchange and quality assurance, making it more successful than most other forms of closed or open peer review:

- Publication of discussion papers before full peer review and revision: free speech, rapid publication, and public accountability of authors for their original manuscript foster innovation and deter careless submissions.
- Integration of public peer review and interactive discussion prior to final publication: attract more comments than post-peer review commenting, enhance efficiency, and transparency of quality assurance, maximize information density of final papers.
- Optional anonymity for designated referees: enables critical comments and questions by referees who might be reluctant to risk appearing ignorant or disrespectful – especially when providing a voluntary community service in which they have little to gain for investing lots of effort and time.
- Archiving, public accessibility, and citability of every discussion paper and interactive comment: ensure documentation of controversial scientific innovations or flaws, public recognition of commentators' contributions, and deterrence of careless submissions.

Combining all of the above features and effects is the basis for the great success of ACP and its sister journals. Missing out on one or more of these features is the main reason why most if not all alternative forms of peer review practised in other initiatives for improving scientific communication and quality assurance have been less successful (less commenting, lower impact/visibility, higher rejection rates, larger waste of refereeing capacities, etc.).

For example, the release of a "pre-publication history" and/or the opportunity for "peer commentary" after completion of the actual peer review and publication of the final revised manuscript as practiced by the BMC medical journals of BioMed Central[29] as well as the journals Behavioral and Brain Sciences[30] and Psychology[31] are very useful advances and improvements compared to traditional journal publishing, but they miss some of the above features and advantages. Controversial scientific innovations or flaws in papers rejected after peer review are not documented for the public and scientific community. Moreover, the completion of peer review and revision before publication and public discussion of a manuscript does not allow interested members of the scientific community to have any input to the revision and the final editorial decision. Obviously, "post-commenting" after peer review is much less attractive to scientists than commenting in the course of peer review. The latter allows individual scientists to support and influence the conclusions and publications of their colleagues, e.g., by pointing out related earlier findings and studies which the authors can still include in the reference list of the manuscript thus in standard citation analyses. In contrast, post-commenting after final publication does neither enable the commentator to influence the final publication, nor does it allow the authors to improve their publication along the lines suggested by the commentators. Accordingly, potential commentators have not only less incentive to invest effort and time in contributing to their colleagues' and competitors' work; they also have to worry that critical comments might just be regarded as a devaluing critique rather than a helpful contribution. This fairly straightforward consideration is supported by the fact that most journals with post-commenting receive fewer comments from the scientific community (Müller, 2008). For example, only one of ~20 papers published in PLoS One receives a comment from the scientific community (as opposed to one of ~5 in ACP), although PLoS offers more advanced and easier to use commenting tools and tries to advertise and promote the commenting more actively than ACP.

For several reasons also the "open peer review trial" of the Nature magazine in 2006 was not a good example and measure for the engagement of scientists in interactive commenting and public peer review on the internet. In that experiment, neither the authors of an article nor their colleagues and readers had much of an incentive to participate in the public discussion. The authors had to accept that their article was exposed in parallel to public scrutiny as well as to a closed peer review process where the referee comments remain non-public and where most of submitted manuscripts are rejected not because of a lack of scientific quality but because they are not deemed sufficiently exciting for the interdisciplinary audience of the magazine (ca. 93% rejection rate)[32]. For the likely outcome that a manuscript would not pass the closed peer review, it was not clear whether and in which form the rejected manuscript and the public comments would remain publicly accessible. As one might have imagined beforehand, this is not a very attractive perspective for scientists trying to get recognition for their most exciting results. Similarly, colleagues and readers had little incentive to formulate and post substantial comments, because their contributions would just have been an addendum to the closed peer review proceeding in parallel and would likely disappear afterward. Fortunately, the publishers of Nature seem to have realized that permanent archiving and citability are key features of scientific exchange, and they have launched a more promising initiative titled Nature Precedings. There manuscripts can be published, openly discussed and archived in a similar way as in the discussion forums of interactive open access journals[33].

Unfortunately, however, it seems that the paramount importance of archiving and citability of manuscripts and comments has not yet been fully recognized by scientific publishers and societies. Following up on the success and leadership of the EGU in interactive open access publishing and peer review, the American Geophysical Union (AGU) has recently also engaged in experiments with "open peer review." Instead of building on the very positive experience and success of the European sister society, however, AGU seems to follow the tracks of the unsuccessful earlier trial of Nature. Specifically, AGU announced that the discussion paper and all interactive comments shall be deleted after completion of the peer review process and final acceptance or rejection of the revised manuscript (Albarede, 2009). This line was also followed in the JAMES, which had originally adopted the interactive open access journal concept of ACP but then abandoned the archiving of discussion papers in their discussion forum (JAMES-D) and was recently taken over by AGU. If AGU were to continue the approach of erasing discussion papers and comments, they would largely miss out on the effects detailed under point 4 above, and it appears questionable that the perspective of deletion after a couple of months will attract substantial commenting from the scientific community. Hopefully, the proponents of the AGU experiment will realize that the deletion of scientific comments is not only a discouragement for potential commentators but also a regrettable underestimation of the value of scientific discussion and discourse in the history and progress of science.

As outlined on the web pages of ACP/EGU, the permanent archiving of discussion papers can occasionally lead to inconveniences for authors and other parties involved in the review and publication process. Overall, however, the advantages of permanent archiving clearly outweigh the potential disadvantages[34, 35].

For the following reasons it would be neither appropriate nor possible to delete discussion papers after they have been published online:

- The deletion of published materials is incompatible with the virtues of traceability and reliability that are central to science and scientific publishing in general, and to the interactive open access publishing approach of ACP/EGU in particular. Deleting published scientific information is against the very nature of science.

- The deletion of discussion papers and comments would discourage potential commentators, and it would imply a disregard for the value of scientific discourse (Pöschl, 2010b, pp. 305–306).

- The use of digital object identifiers (DOI) entails legal obligations of ensure permanent archiving and accessibility.

- Even if it were desirable, it would be practically impossible to "unpublish" a discussion paper published in ACPD. Upon online publication, the papers are copied into multiple electronic repositories. Moreover, referees, readers, and other internet users can and do download copies for storage at arbitrary locations that are beyond the control of any the publisher. Therefore, a published paper can be formally withdrawn/retracted by publication of a commentary analogous to stating the reasons like in traditional print journals. It can, however, not be "unpublished" by deletion from the web pages and archives of the journal.

One of the central aims of interactive open access publishing is high efficiency in scientific communication and quality assurance. As detailed in the attached articles, the average quality and visibility of ACP, and its sister journals are higher than those of most comparable journals while the rejection rates are lower. The highly efficient mechanism of review, publication, and self-regulation would hardly work if authors could submit manuscripts at any rate and simply delete published discussion papers if the public peer review and editorial decision were not favorable (or for any other reason).

Experience and rational thinking suggest that multi-stage open peer review should be applicable and beneficial for journal publications in most if not all disciplines of scientific research (STM as well as social sciences, economics, and humanities). For consistency and traceability, discussion papers, and interactive comments should generally remain archived and citable as published, and they should be regarded as proceedings-type publications. Due to the proceedings character of discussion papers, the authors of revised manuscripts that may not have been accepted for final publication in the interactive open access journal to which they had originally been submitted can still pursue review and publication in alternative journals. As indicated above, such aspects are particularly important with regard to highlight magazines or journals in which the review process is not only aimed at ensuring scientific quality but also at high selectivity with regard to interdisciplinary relevance and visibility, which entails low probability of acceptance even for manuscripts of high quality (see Nature trial).

In addition to the above general features, the following specific procedural aspects have turned out to be important for the practical implementation and effectiveness of interactive open access publishing and peer review:

EDITOR ASSIGNMENT

For the assignment of a newly submitted manuscript to a handling editor, the online editorial office automatically sends invitation letters to all editorial board members covering the relevant subject areas (based on index terms selected by authors). Depending on competence and availability, each editorial board member can then decide if s/he wants to take editorship (first come, first served; every board member is expected to handle at least six submissions per year). If no handling editor can be found via the automated assignment process, the authors and the executive editors are informed and asked to directly contact individual board members if they are ready to take editorship. This second line of editor assignment in ACP is similar to the regular editor assignment procedure in the open access journal Biology Direct[36].

There it is up to the authors to find and motivate an editorial board member to guide the review process for their

ACCESS REVIEW

Prior to publication in the discussion forum, the editor is asked to evaluate whether the submitted manuscript is within the scope of the journal and whether it meets basic quality criteria. If necessary, the editor may consult referees for a rapid and preliminary initial rating of the manuscript[37]. The editor or referees can request/suggest minor technical corrections and adjustment (typing errors, clarifications, etc.). Further requests for revision of the scientific contents are not allowed at this stage of the review process but shall be expressed in the interactive discussion following publication of the discussion paper. For rapid processing and in order to save refereeing capacities the editor shall normally perform the access review without the referees, unless their advice is urgently needed or the authors have requested their involvement. In a statement or cover letter accompanying the submitted manuscript, the authors can indicate if they have any preference on involving the referees already in the access review. Obviously, the involvement of referees can lead to delays, but on the other hand the authors may want to receive a preliminary rating and suggestions for minor corrections prior to publication of the discussion paper.

FINAL RESPONSE AND REVIEW COMPLETION

In the final response phase at the end of the interactive public discussion, the authors shall respond to all comments. The editor has the opportunity of adding comments and suggestions, but normally editorial decisions and recommendations should not be taken and expressed before the authors have responded to all comments ("audiatur et altera pars"). Instead, it shall be up to the authors to decide if they want to pursue final publication and how they shall revise their manuscript in view of the public review and discussion (self-regulation once again). Depending on the situation, they can but need not ask and wait for the editor to give advice on how to proceed and if a revised version is likely to be accepted for final publication. After receiving critical feedback, mature, and responsible scientists should normally know best how to revise their manuscript. Indeed, the improvements upon revision of a manuscript after public discussion often go far beyond the requests and suggestions expressed by the referees. Premature interference by the editor would likely reduce rather than enhance the authors' motivation for improving the manuscript upon revision. Moreover, premature editorial recommendations published by the editor before seeing the authors' final response and the revised manuscript could potentially bias the final decision about acceptance or rejection.

After receiving the revised manuscript the editor has a complete picture, can check if all comments and suggestions have been properly taken into account, and can suggest or request further improvements. If required, the process of review and revision can be iterated with the help of referees. So far, such iterations of peer review as well as appeal procedures in case of controversial editorial decisions have not been handled in public to avoid unnecessary complications. In the end, however, the discussion forum can and shall be used to explain editorial decisions in a rational and transparent way as illustrated by the following example (Pöschl, 2009b)[38]:

Currently, the editorial guidelines of ACP encourage editors to publish scientifically useful referee-author exchanges from non-public part of peer review completion in similar ways as in the exemplary case cited above. In the future, intermediate manuscript versions and related comments from the access review or the review completion shall be automatically made available upon publication of a manuscript in ACPD or ACP, respectively (analogous to pre-publication history available in BMC medical journals). If, however, a newly submitted manuscript is not accepted for publication in ACPD or a revised manuscript is not accepted for publication in ACP, the manuscript, and related comments shall be kept confidential in order to avoid escalation of scientific disputes and to maintain the authors' opportunity of pursuing publication in alternative publishing venues (European Geosciences Union, 2010).

COMPARISON TO EARLIER INITIATIVES WITH TWO- OR MULTI-STAGE OPEN PEER REVIEW

Following up on the requests of a referee in the peer review of this manuscript, the following paragraphs provide a detailed comparison to earlier initiatives with similar concepts and a discussion of potential reasons for different developments. During the initiation and planning of ACP and its interactive journal concept in the years 2000 and 2001, I was looking for – but was unable to find – similar initiatives to compare with and learn from (Pöschl, 2004). It was only at an e-publishing workshop of the Max Planck Center for Information Management in May 2002 that I learned of a similar initiative launched as early as 1996: the JIME[39]. Coming from a completely different scientific background, the founders of JIME had designed and realized a similar concept of multi-stage open peer review with public discussion. Unfortunately, however, JIME attracted only a small number of publications and seems not to have inspired the foundation of similar journals in related fields of science and humanities. Despite the overall conceptual similarities, JIME does not show some of the key features of the ACP interactive journal concept. In particular, the "private open peer review" of JIME foresees a non-public exchange of arguments between referees and authors, which is opened to the public only after approval by the editor. This seems to limit the publication and documentation of controversial scientific innovations or flaws much more than the "access peer review" of ACP (quick go/no-go decision essentially without non-public exchange of arguments between authors and referees). Moreover, all referees are named and no anonymous referee comments are allowed in JIME, which is likely to limit and inhibit critical review and discussion. These differences may appear subtle at first sight, but they are highly relevant for the practical operation of a scientific journal and may be decisive for its success and acceptance in the target scientific community.

After JIME, I got to know about another early online publication format with a two-stage open peer review process: the ETAI[40] launched in 1997. Similar to JIME, ETAI attracted a series of special issues related to conferences or projects, but the number of individually submitted articles remained small. Regular operations stopped in 2002, but the ETAI home page indicates plans for a re-launch. As described by Sandewall (1997, 2006, 2012), the open peer review process of ETAI does not integrate but separate the two major aims of peer review, namely, to improve the quality of submitted manuscripts and to establish certain quality standards. The first stage is an interactive public discussion which invites questions, comments, and suggestions from the scientific community, but it does not involve designated referees, and all participants are openly named. In a second stage, anonymous referees decide about acceptance of the revised manuscript for ETAI, and further rounds of revision are normally not allowed. These features of ETAI bear similarities to the unsuccessful trial of open peer review by Nature magazine in 2006, and they are in stark contrast to the ACP review process, where the referees contribute to the interactive public discussion and have an option of staying anonymous, and the peer review process can be continued iteratively like in traditional journals. For the authors and readers of ETAI it seems not clear, if the openly named participants of the interactive public discussion in the first stage of the review process might also serve as an anonymous referees in the second stage. It seems rather unattractive for authors to post their manuscript for open discussion and scrutiny by the scientific community, and to have only one chance of revision before anonymous referees who may or may not have been involved in the preceding discussion are expected to make a "pass/fail decision" (Sandewall, 2012). In the relatively few review processes that have actually been completed in ETAI so far (several dozens in the time frame of 1997–2002), all involved parties seem to have requested exceptions, i.e., anonymity in the interactive public discussion and iterative revisions in the second stage of review (Sandewall, 2012). Both of these "exceptional" features are key elements of the successful ACP approach. From long-term experience with several thousand review processes completed in ACP since 2001, we know that these features are vital for the large success of the EGU interactive open access journals, and I would argue that they might be critical for the limited success of ETAI. In any case, the ACP/EGU approach of multi-stage open peer review is aimed at integrating rather than separating the processes of interactive public discussion and classical peer review as well as the aims of manuscript improvement and quality control.

The limited success of JIME and ETAI compared to ACP demonstrates the difficulties of practical implementation and the importance of the conceptual aspects and subtleties outlined above (see Key features of multi-stage open peer review as practiced by ACP). Nevertheless, the basic aims and principles of JIME, ETAI, and ACP are similar, and their independent development in different disciplines including the social, natural, and computer sciences reflects the power of the idea and the appeal of transparency in scientific quality assurance.

The review article of Sandewall (2012) outlines and compares further analogies and differences between ETAI and ACP, and it also provides a very useful and comprehensive account of challenges faced by proponents of open peer review. In the following paragraphs I am following up on some of the questions and issues raised.

- Defining different types of scientific publication (Sandewall, 2012: p. 2–3): Robust and self-consistent definitions of different types of scientific publications are indeed important for scientific communication and quality assurance. I would, however, not tie such definitions to electronic vs. non-electronic or different types of publishers. Instead, I would suggest to use self-explanatory terms that are meaningful regardless of the publishing medium. Along these lines, the term "discussion paper" has proven well defined and useful as specified in a position statement of the EGU with references to other scientific societies and publishers. Thus, I would recommend broad usage of this term for the first stage of publication in two- or multi-stage open peer review.

- Resolving doubts about the viability of open peer review (Sandewall, 2012: Section 4.1): For the reasons outlined by Sandewall (2012) it is important to demonstrate the viability and advantages of open peer review with practical examples. The statistics of ACP and its sister journals prove that the arguments given in Section 4.1 of Sandewall (2012) are valid and applicable to a wide range of research areas involving scientists trained in physics, chemistry, biology, geology, engineering, and other disciplines. Besides a clear concept and terminology ("discussion paper," etc.), it is important to have a dedicated team of scientists who do not only advertise and explain the new approach but also demonstrate its practical viability by submitting and publishing high quality papers (see below).

- Starting the flow of submissions and debate (Sandewall, 2012: Section 4.2): Starting a steady flow of submissions is indeed the most important task for the editorial board of any new journal – even more so for an innovative journal experimenting with new forms of peer review. In most areas of natural science, a journal can be regarded as well established only when it is covered by major indexing services and acquires a journal impact factor or equivalent measure of visibility, which usually takes at least a couple of years. Until then, colleagues without genuine interest in the journal cannot be expected to submit high quality manuscripts that would likely reach higher visibility and citation counts elsewhere. Thus, it is up to the editorial board members and other supporters to maintain a steady flow of high quality submissions. For this purpose as well as for efficient handling of manuscripts when the flow of submissions increases, it is helpful to gather a large editorial board that is firmly rooted in the scientific community and includes experts for all subject areas of the journal scope (ACP: ~70 board members at the beginning, ~130 now).

- Initiating the review and discussion of manuscripts with high quality comments that set a precedent for further commenting is of course also important for journals with open peer review. In ACP and its interactive open access sister journals this is mostly done by designated referees appointed by the editor handling the submission. Unsolicited comments can be expected only if members of the scientific community have a strong interest to ask for more information or suggest corrections/additions concerning the methods, results, and conclusions of a study. As expected, non-controversial papers usually receive comments only from the designated referees. Other scientists have little incentive to invest effort and time commenting on papers that they may find potentially useful but not controversial.

- Maintaining coherence (Sandewall, 2012: Section 4.3): For ACP, coherence is not more of an issue than for traditional journals covering multiple subject areas with the help of multiple editors. The journal scope has to be well defined and reflect the interests and quality standards of the scientific community served by the journal. Different communities tend to have different standards and preferences with regard to both the format and the content of manuscripts. Therefore, EGU publishes multiple topical journals rather than just one large geosciences journal including all disciplines. Even within the discipline of atmospheric science, EGU publishes more than just one journal, namely ACP and the sister journal "Atmospheric Measurement Techniques" (AMT) which is focused on method development and exhibits similarly high growth rates of volume and visibility as ACP. Due to the transparency of the review process and related self-regulation mechanisms, the quality of final papers published in ACP is generally not more variable than in traditional journals with smaller editorial boards. The ACP editors do not spend extra time on moderating the interactive public discussions, which are not actively moderated for the reasons outlined above. Compared to traditional journals where the editors often rely on simple majority votes of the referees, however, the ACP editors tend to spend more time on carefully validating the referee recommendations, because the transparent review process publicly reveals editorial decisions that are not well-founded.

- Computational and administrative infrastructure (Sandewall, 2012: Section 4.4): The installation and maintenance of computational and administrative infrastructure is the main reason why the operation of an open access journal is not cost free, even if most of the review work is done by volunteers. The referees and editors of EGU journals receive no financial rewards. The editors even pay the regular registration fee to participate in the annual EGU General Assemblies with over 10,000 participants where the editorial board meetings take place. The small commercial publisher Copernicus is a spin-off from the Max Planck Society and continues to aim for providing optimal infrastructure and services at minimal cost. Nevertheless, it seems difficult to reduce the average costs far below one thousand Euros per paper, but this is anyhow much lower than the prices of most traditional publishers as discussed above (see Financing and Sustainability of Interactive Open Access Publishing).

- Maintaining liveliness of peer review discussion (Sandewall, 2012: see Comparison to Earlier Initiatives with Two- or Multi-Stage Open Peer Review): For the reasons outlined by Sandewall (2012), it is difficult if not impossible to ensure a lively review discussion for all papers published in large scientific journals.

This may be problematic for the two-stage review approach of ETAI, where the first stage is designed as a pure community discussion without the involvement of designated referees. For the integrative approach of ACP, however, it is not problematic that most papers receive comments only from the designated referees. The transparency of the peer review process and the option for additional input from the scientific community are sufficient to stimulate self-regulation and enhance the efficiency of scientific quality assurance (Pöschl, 2004, 2010a,b). Discussion papers that report controversial findings often do attract unsolicited comments from the scientific community, but why would researchers invest effort and time in the commenting of their colleagues' publications which they may find interesting but not controversial? Sometimes more commenting and discussion might be useful, but usually the volume of comments exchanged between authors and referees amounts to as much as 50% of the discussion paper volume, and further commenting can be cumbersome – especially for the authors who normally do not want to spend too much time and effort on the discussion of a single paper but rather move on to the next study. Therefore, unnecessary comments and artificial liveliness of discussion might actually deter authors and do more harm than good to a journal with open peer review.

- Open names policy (Sandewall, 2012: Section 7.1): In an ideal world, where people generally react positive to criticism and where scientists can dedicate unlimited amounts of time and effort into compiling completely accurate reviews about their colleagues' manuscripts, I would agree that referee anonymity should be abandoned. In practice, however, optional anonymity for referees appears appropriate or even necessary to enable critical comments and questions by referees who might be reluctant to risk appearing ignorant or disrespectful (Pöschl, 2004). As outlined above, less than 20% of the referee comments published in the discussion forum of ACP are posted with the name of the referee, i.e., the referees prefer in most cases (>80%) not to reveal their name. Purists often suggest that offering anonymity to referees would be unfair against the authors of a manuscript, and that both parties should be openly named to ensure equal rights and opportunities. They tend to forget, however, that the authors want to get their paper approved by peers, and that the referees usually provide this service on a voluntary basis. In this sense, the authors actually exploit the working capacities of the referees, and the peer review process offers a major gain to the authors (conversion of their manuscript into a peer reviewed paper) but relatively little benefit to the referees. Therefore, it seems appropriate to protect the referees from potential negative consequences of the free service they provided to the authors and to the scientific community. The very small number of author complaints about inappropriate referee comments (about one in 10,000) and the low rejection rates of manuscripts submitted for peer review in ACP and the other EGU interactive open access sister journals (generally <15%) confirm that transparency of the review process (open-process peer review) is normally sufficient to protect authors from inappropriate referee comments. Thus, it seems neither necessary nor appropriate to abandon optional anonymity, impose an open names policy and force referees to reveal their identity. All available evidence suggests that refereeing capacities are the most limited resource in scientific publishing and quality assurance (Pöschl, 2004). In view of the ever-increasing flow of manuscripts submitted for peer reviewed publication, it appears more important to protect referees rather than authors – especially in a multi-stage open peer review process like that of ACP, where the authors anyhow have the opportunity of free speech through their discussion paper and the interactive comments they can post during the open discussion as well as in a final response phase where no more referee comments are allowed[41].

KEY QUESTIONS FOR OPEN EVALUATION IN SCIENCE

The coordinators of the special issue hosting this article posed a series of ten key questions to be considered in designing and implementing a concept of open evaluation in science. More than a decade of practical experience and success in re-shaping the processes of scientific publishing and quality assurance as well as continued exchange with scientists and publishing professionals from various disciplines in the sciences and humanities lead to the following answers.

- Should some evaluation take place prior to publication or should all evaluation occur post-publication? Experience and rational consideration suggest that the main review process should take place before (final) publication of a manuscript. A fundamental disadvantage of pure post-publication review is that the reviewers cannot contribute to a revision and improvement of the published manuscript. Thus, both the authors and the reviewers are likely to consider critical comments as destructive rather than constructive. Moreover, the reviewer has less incentive to invest effort and time in suggesting additions and corrections, including but not limited to referencing relevant related publications. Last but not least, post-publication commenting does not enhance the information density of scientific communication. If the reviewer comments cannot be implemented in a revised manuscript, the readers have to consult all comments and extract the information from there, which is much less efficient than reading a revised manuscript that synthesizes the information exchanged in the review process. For the above reasons, most publishing platforms that offer only post-publication commenting attract rather small numbers and volumes of comments.

- Should reviews and ratings be entirely transparent, or should some aspects be kept secret? Reviews and ratings pertaining to a published manuscript should be made entirely transparent. Reviews and ratings of manuscripts that do not achieve (final) publication, however, should be kept confidential to avoid public escalation of scientific disputes and to give authors a chance of pursuing publication of their (revised) manuscript in alternative publishing venues.

- Should alternative metrics, such as paper downloads be included in the evaluation? Paper download statistics are among the many possible forms of post-publication evaluation and should certainly be considered for comprehensive evaluation of scientific publications, but not without precautions against manipulation and misinterpretation of this relatively primitive usage metric. Many scientific journals, including traditional subscription journals with hidden peer review, are already providing download data and highlighting most downloaded papers. This approach certainly facilitates the detection of "hot papers," but compared to long-term citation statistics and other usage metrics it seems less robust and should not be overrated.

- How can scientific objectivity be strengthened and political motivations weakened in the future system? Like in all branches of human society and politics, transparency, and free speech appear to be the best if not the only sustainable way of pursuing objectivity in a balance of powers and interests.

- Should the system use signed and authenticated reviews and ratings or anonymous ones, or both? An entirely open and traceable exchange of scientific arguments in the form of signed and authenticated comments is certainly desirable and shall be encouraged. For practical reasons, however, it seems appropriate and beneficial to allow also for anonymous reviews. Optional anonymity enables critical comments and questions by referees who might be reluctant to risk appearing ignorant or disrespectful – especially when providing a voluntary community service in which they have little to gain for investing lots of effort and time.

- Should the evaluation be an ongoing process, such that promising papers are more deeply evaluated? The evaluation of scientific publications has to be and generally is an ongoing process – with regard to citation counting as well as commenting and other forms of evaluation that are and have long been in use. Note that also traditional journals with hidden peer review also allow for commentaries referring to earlier papers. In practice, however, relatively few papers seem to attract comments after (final) publication. Moreover, most authors seem to prefer finalizing a publication at some point, and following up with new studies rather than continuously revising and updating old papers. For certain types of publications such as review articles, continuous extension, and revision may be a good and attractive approach as exemplified by the Living Reviews project and journal family[42]. For standard articles presenting new scientific findings, however, a finite process of publication appears more straightforward. Either way, thorough evaluation of scientific studies seems difficult if not impossible without long-term perspective.

- How can we bring science and statistics to the evaluation process (e.g., should rating averages come with error bars)? Scientific reviews and ratings are necessarily subject to the same uncertainties and progress as the studies that undergo rating and review. Thus, it seems natural to assess also the reliability of reviews and ratings. One of the many advantages of open peer review is the public availability of reviews and ratings, which makes them accessibly for statistical analysis. Thus open access and open peer review inherently promote the development of new and improved evaluation metrics – in analogy to traditional indexing services like the ISI Web of Science and Elsevier's SCOPUS, but much more efficiently and comprehensively because of unrestricted access and free competition for optimal solutions.

- How should the evaluative information about each paper (e.g., peer ratings) be combined to prioritize the literature? The combination and balancing of different types of evaluative information (ratings/reviews, download/citation statistics, and other usage metrics) will necessarily depend on the aims and perspectives of different types of evaluation or prioritization. For example, the criteria of an evaluation exercise will likely differ for individuals and institutions, scientific researchers and teachers, innovation, and reliability, short-term and long-term impact, etc. In any case, it should be kept in mind statistical indicators are sometimes useful but always also prone to misinterpretation (see publication and citation counting, impact factors, h-indices, etc.).

- Should different individuals and organizations be able to define their own evaluation formulae (e.g., weighting ratings according to different criteria)? Obviously, different individuals and institutions may pursue different goals and should thus be able to apply different criteria and weighting schemes. Moreover, evaluators and service providers should compete in developing the best possible metrics and indicators. This is already the case with ISI Web of Science and Elsevier SCOPUS, and through open access and open peer review many more parties can participate, contribute, and help to overcome the obsolete monopoly/ oligopoly structures of scientific indexing.

- How can we efficiently transition toward the future system? An efficient transition to open evaluation in science can be achieved by combining the strengths of traditional peer review with the opportunities of interactive and transparent community assessment on the internet. The concept of multi-stage open peer review has been designed and successfully applied to induce this transition in the geosciences and is spreading into other disciplines. It can be flexibly adjusted to the needs and peculiarities of different scientific communities, and it has the potential of replacing hidden peer review as the standard of scientific quality assurance and forming the basis of an open evaluation system.

CONCLUSIONS AND OUTLOOK

ACP and its sister journals very clearly demonstrate that interactive open access publishing with a multi-stage peer review process effectively resolves the dilemma between rapid scientific exchange and thorough quality assurance. They have proven that multi-stage open peer review indeed fosters scientific discussion, deters submission of sub-standard manuscripts, saves refereeing capacities, and enhances information density in final papers. Moreover, ACP, EGU and Copernicus prove the financial sustainability of open access publishing, and they may serve as a role model for how STM publishing at large can manage the transition from the past of print-based subscription barriers into the future of internet-based open access. The key for a successful, smooth, and efficient transition is to utilize the opportunities of modern technology and interactivity while maintaining the strengths of traditional structures and procedures.

Multi-stage open peer review easily can be integrated into new and existing scientific journals as well as large-scale publishing systems and repositories such as arXiv.org – simply by adding an interactive discussion forum. Equipped with appropriate interactive commenting tools, a large repository such as arXiv.org could not only serve as an archive for "preprints" or "e-prints," but also as a platform for efficient review and discussion, where authors could post their discussion papers and different journals could send their referees for public review. Similarly, individual publishers could set up central discussion forums to serve different journals or journal sections (Pöschl, 2004,2010b). This perspective is in line with the selected papers network concept of Lee (2012) and the decoupled journal concept of Priem and Hemminger (2012). Depending on the outcome of public review and discussion, the revised manuscripts could then be sorted and grouped at different levels of relevance for different audiences – analogous to the quality ranking system and tiers of the Berkeley Electronic Press journals in economics[43],[44]. Another feature that could be integrated in multi-stage open peer review is a double-blind approach in the initial access review (pre-screening) to avoid/minimize bias in selection of discussion papers. In the open discussion, however, it seems more useful and efficient to discuss openly without hiding identities (except for protecting referees if they wish to stay anonymous).

For interdisciplinary highlight papers, EGU and Copernicus are currently preparing the introduction of a third stage of interactive open access publishing that shall lead to efficient grouping of scientific publications in three tiers with the following characteristics:

- Discussion forum (discussion papers and interactive comments):
  1. free speech (for authors and scientific community)
  2. original opinions
  3. immediate publication and dissemination
- Topical journal (final papers):
  1. thorough quality assurance (collaborative peer review)
  2. comprehensive, complete and validated information
- Highlight magazine (abstracts):
  1. highly condensed information
  2. interdisciplinary relevance and public interest
  3. three-stage selection process (distillation).

The interactive open access highlight magazine shall be dedicated to the selection and presentation of the abstracts of highlight papers, which outline the forefront of research and are of high interdisciplinary relevance and public interest. The editorial board of the magazine shall select highlight papers that have undergone public peer review and discussion in topical open access journals, and the abstracts of the highlight papers shall be commented and compiled with direct references and links to the original papers and journals, respectively. By building on rather than competing with topical scientific journals, the highlight selection process and magazine shall provide high efficiency, conciseness and interdisciplinarity without compromising scientific completeness and quality assurance. This might also be a way forward for traditional highlight magazines like Nature or Science covering the full width of scientific disciplines.

The basic concepts of interactive open access publishing and peer review can be easily adjusted to the different needs and capacities of different scientific communities by maintaining or abandoning referee anonymity, shortening, or prolonging the public discussion phase, adding post-peer review commenting and rating tools for readers, making all steps/iterations of peer-review and revision transparent, adding further stages of publication for re-revised manuscripts, establishing feedback loops for editorial quality assurance, etc.

Figure 3 illustrates essential elements and scales of evaluation in an open system of scientific publication and quality assurance based on multi-stage open peer review. While much of the general discussion about reforming scientific quality assurance and evaluation is focused on a distinction of pre- and post-publication processes, the experience and achievements of ACP and EGU show that an integrative approach combining pre- and post-publication elements in a multi-step process of review and publication is most efficient.

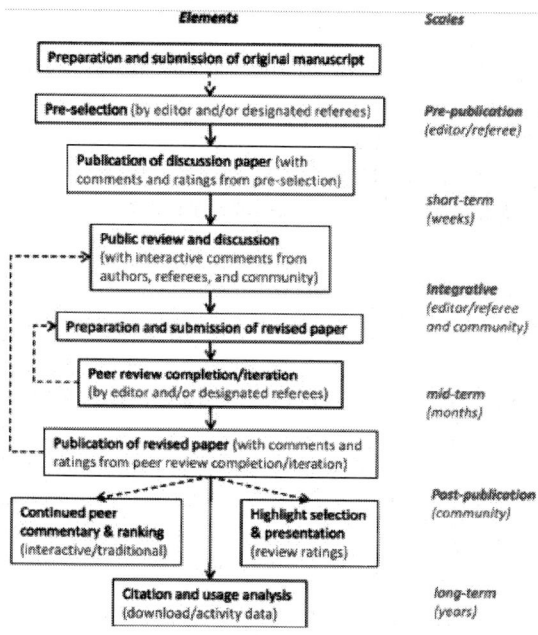

Figure 3: Elements and scales of evaluation in an open system of scientific publication and quality assurance based on multi-stage open peer review. Solid and dashed arrows indicate essential and optional processes.

Besides communication and evaluation of scientific results, multi-stage open peer review might also be applicable for efficient evaluation of scientific research proposals in the form of citable discussion papers. Again all involved parties could profit from public documentation, scrutiny, and citability. At first sight, it might appear that the authors of a proposal would run a high risk of "losing" innovative project ideas to the public. In practice, however, they might be better protected from (hidden) plagiarism and obstruction by competitors, and the citable publication might actually help them to claim authorship, precedence, and recognition for their ideas. At the same time, the scientific community and society at large might profit from rapid dissemination of innovative ideas.

Overall, interactive open access publishing and peer review can strongly enhance scientific exchange and quality assurance. The concept has been very successfully applied and extended over the past decade, demonstrating both the scientific benefits and the financial sustainability of open access. It will likely emerge as a best practice model for the future of scientific publishing, and it provides a solid basis for efficient use and augmentation of scientific knowledge in the global information commons (David and Uhlir, 2005). Moreover, public review, discussion, and documentation of the scientific discourse can serve as an example for rational and transparent procedures of settling complex questions, problems, and disputes. It is a model for further development of the structures, mechanisms, and processes of communication and decision making in society and politics in line with the principles of critical rationalism and open societies.

A major limiting factor for the development of innovative scientific publication and evaluation systems is the scarcity of funds specifically dedicated to covering open access publication costs. Nevertheless, more and more funding agencies do provide funds for this purpose, and the success of the EGU/Copernicus as well as other open access publishers shows that many scientists are willing and able to cover the costs of open access publishing via publication fees. Overall, the money required to produce scientific publications in a format that is accessible via the internet is already in circulation. Otherwise, the publishers would not be able to offer online subscriptions. Currently, however, the funds are channeled through a rigid subscription system, which has the consequence that certain publishers can make excessively large profits and that the scientific information remains locked away.

If the same amount of money were channeled through a flexible open access funding schemed, the same products (scientific journals and papers) could be produced and made freely available on the internet at the same or lower cost in a proper publishing market rather than the current subscription scheme with oligopoly character.

In order to accelerate the improvement of scientific communication and evaluation in a global information commons, I would like to renew the following propositions and recommendations to scientists and scientific publishers, librarians, institutions, and funding agencies (Pöschl, 2004,2010b):

Promote open access to publicly funded research publications by appropriate guidelines and by moving funds from subscription budgets to publication budgets – preferably at high rates (20% per year or more). Obviously, traditional publishers are reluctant to undermine their profits as long as they can rely on rigid subscription schemes, but the ones who are ready to serve science will swiftly adapt to new financing schemes as illustrated by the open choice model and acquisition of BioMedCentral by Springer[45]. The others can be substituted by new service providers as indicated by the swiftly growing number, size, and visibility of open access publishers and journals[46,47].

- Promote multi-stage open peer review in new and existing journals, repositories, and other publication platforms. Public review and interactive discussion are technically straightforward and can be flexibly adjusted to different scientific communities, but care should be taken when dealing with key features of peer review and scientific discourse (optional anonymity for designated referees, permanent archiving, and citability of published manuscripts and comments, etc.).

- Promote the development and implementation of new and improved metrics for the impact and quality of scientific publications (combination of citation, download/usage, commenting, and ranking by various groups of readers and users, respectively). Note that open access is urgently needed to stimulate innovation by competition in this field, which has long been hampered by monopoly structures. The working capacities of librarians and related information professionals that may be liberated by the end of the subscription business are urgently needed for the structuring, processing, quality assurance, and digital preservation of scientific contents, bibliometric data, and statistical analyses both at scientific institutions and at commercial service providers.

ACKNOWLEDGMENTS

The author thanks A. Pöschl, M. Pöschl, and M. Weller for inspiration and support. P.J. Crutzen, T. Koop, K.S. Carslaw, R. Sander, W.T. Sturges, A.K. Richter, M. Rasmussen, the scientific communities of ACP and EGU, and the team of Copernicus are gratefully acknowledged for successful and enjoyable collaboration in the development of interactive open access publishing and multi-stage open peer review. Three referees are gratefully acknowledged for stimulating comments in the peer review of this paper.

REFERENCES

1. Albarede, F. (2009). AGU announces open peer review experiment. *Eos (Washington DC)* 90, 276.

2. Bodenschatz, E., and Pöschl, U. (2008). "Open access and quality assurance," in *Open Access Challenges and Perspectives – A Handbook, European Commission and German Commission for UNESCO.* Available at:http://ec.europa.eu/research/science-society/document_library/pdf_06/open-access-handbook_en.pdf

3. Bollen, J., Van de Sompel, H., Hagberg, A., Bettencourt, L., Chute, R., Rodriguez, M. A., and Balakireva, L. (2009). Clickstream data yields high-resolution maps of science. *PLoS ONE* 4, e4803. doi:10.1371/journal.pone.0004803

4. Bornmann, L., and Daniel, H. D. (2010a). Reliability of reviewers' ratings when using public peer review: a case study.*Learn. Publ.* 23, 124–131.

5. Bornmann, L., and Daniel, H.-D. (2010b). Do author-suggested reviewers rate submissions more favorably than editor-suggested reviewers? A study on atmospheric chemistry and physics. *PLoS ONE* 5, e13345. doi:10.1371/journal.pone.0013345

6. Bornmann, L., Marx, W., Schier, H., Thor, A., and Daniel, H.-D. (2010). From black box to white box at open access journals:

predictive validity of manuscript reviewing and editorial decisions at atmospheric chemistry and physics. *Res. Eval.*19, 105–118.

7.   Bornmann, L., Neuhaus, C., and Daniel, H.-D. (2011a). The effect of a two-stage publication process on the Journal Impact Factor: a case study on the interactive open access journal atmospheric chemistry and physics. *Scientometrics* 86, 93–97.

8.   Bornmann, L., Schier, H., Marx, W., and Daniel, H.-D. (2011b). Is interactive open access publishing able to identify high-impact submissions? A study on the predictive validity of atmospheric chemistry and physics by using percentile rank classes. *J. Am. Soc. Inf. Sci. Technol.* 62, 61–71.

9.   Copernicus. (2011). *A Short History of Interactive Open Access Publishing.* Göttingen: Copernicus Publications.

10.   David, P. A., and Uhlir, P. F. (2005). "Creating the information commons for e-science: toward institutional policies and guidelines for action," *Workshop Proceedings*, (Paris: UNESCO).

11.   EconomistAcademicPublishing.(2011).*OfGoatsandHeadaches.* Available at:http://www.economist.com/node/18744177 [accessed May 26, 2011].

12.   European Commission and German Commission for UNESCO. (2008). *Open Access Opportunities and Challenges – A Handbook.* Available at: http://ec.europa.eu/research/science-society/document_library/pdf_06/open-access-handbook_en.pdf

13.   European Geosciences Union. (2010). *EGU Position Statement on the Status of Discussion Papers Published in EGU Interactive Open Access Journals.* Available at: http://www.egu.eu/statements/position-statement-on-the-status-of-discussion-papers-published-in-egu-interactive-open-access-journals-4-july-2010.html

14.   Golden, M., and Schultz, D. M. (2012). Quantifying the volunteer effort of scientific peer reviewing. *Bull. Am. Meteorol. Soc.*93, 337–345.

15.   Lee, C. (2012). Open peer review by a selected-papers network. *Front. Comput. Neurosci.* 6:1. doi:10.3389/fncom.2012.00001

16.   Max Planck Society. (2003). *Berlin Declaration on Open Access to Knowledge in the Sciences and Humanities.* Available at: http://www.zim.mpg.de/openaccess-berlin/berlin_declaration.pdf

17. Müller, U. (2008). *Peer-Review-Verfahren zur Qualitätssicherung von Open-Access-Zeitschriften – Systematische Klassifikation und empirische Untersuchung*. Ph.D. thesis, Humboldt University, Berlin.

18. Pöschl, U. (2004). Interactive journal concept for improved scientific publishing and quality assurance. *Learn. Pub.* 17, 105–113.

19. Pöschl, U. (2009a). Interactive open access peer review: the atmospheric chemistry and physics model. *Against Grain* 21, 26–34.

20. Pöschl, U. (2009b). Interactive comment on "On the validity of representing hurricanes as Carnot heat engine" by A. M. Makarieva et al. *Atmos. Chem. Phys. Discuss.* 8, S12406–S12411.

21. Pöschl, U. (2010a). Interactive open access publishing and public peer review: the effectiveness of transparency and self-regulation in scientific quality assurance. *IFLA J.* 36, 40–46.

22. Pöschl, U. (2010b). Interactive open access publishing and peer review: the effectiveness and perspectives of transparency and self-regulation in scientific communication and evaluation. *LIBER Q.* 19, 293–314.

23. Pöschl, U. (2010c). *Arne Richter: A multi-talented character who has made a difference in scientific publishing, EGU General Assembly 2010*, Vienna. Available at: http://www.atmospheric-chemistry-and-physics.net/pr_acp_poschl_arne_richter_wien2010_up31.pdf, 2010

24. Pöschl, U. (2011). "On the origin and development of interactive open access publishing," in *A Short History of Interactive Open Access Publishing* (Göttingen: Copernicus Publications).

25. Pöschl, U., and Koop, T. (2008). "Interactive open access publishing and collaborative peer review for improved scientific communication and quality assurance," in *Information Services & Use, 28* (APE 2008: Academic Publishing in Europe, Quality and Publishing, IOS Press), 105–107. Available at: http://www.atmospheric-chemistry-and-physics.net/pr_acp_poeschl_koop_infoservuse_2008_intoapub.pdf

26. Priem, J., and Hemminger, B. M. (2012). Decoupling the scholarly journal. *Front. Comput. Neurosci.* 6:19. doi:10.3389/fncom.2012.00019

27. Sandewall, E. (2006). Opening of the process. A hybrid system of peer review. *Nature.*

28. Sandewall, E. (1997). Publishing and reviewing in the ETAI. *Electron. Trans. Artif. Intell.* 1, 1–12.

29. Sandewall, E. (2012). Maintaining live discussion in two-stage open peer review. *Front. Comput. Neurosci.* 6:9. doi:10.3389/fncom.2012.00009

30. Schultz, D. M. (2010). Rejection rates for journals publishing in the atmospheric sciences. *Bull. Am. Meteorol. Soc.* 231–243.

Chapter 2

Biological and Chemical Diversity of Biogenic Volatile Organic Emissions into the Atmosphere

Alex Guenther

Atmospheric Chemistry Division, NCAR Earth System Laboratory, National Center for Atmospheric Research, 3090 Center Green, Boulder, CO 80301, USA

ABSTRACT

Biogenic volatile organic compounds (BVOC) emitted by terrestrial ecosystems into the atmosphere play an important role in determining atmospheric constituents including the oxidants and aerosols that control air quality and climate. Accurate quantitative estimates of BVOC emissions are needed to understand the processes controlling the earth system and to develop effective air quality and climate management strategies. The large uncertainties associated with BVOC emission estimates must be reduced, but this is challenging due to the large number of compounds and biological sources. The information on the immense biological and chemical diversity of BVOC is reviewed with a focus on observations that have been incorporated into the MEGAN2.1 BVOC emission model. Strategies for improving current BVOC emission modeling approaches by better representations of this diversity are presented. The current gaps in the available data for parameterizing emission models and the priorities for future measurements are discussed.

INTRODUCTION

Terrestrial ecosystems produce and emit many biogenic volatile organic compounds (BVOCs) into the air where they influence the chemistry and composition of the atmosphere including aerosols and oxidants [1–3]. These BVOCs are produced by a variety of sources in terrestrial ecosystems (e.g., flowers, stems, trunks, roots, leaf litter, soil microbes, insects, and animals), but most of the global total emission is from foliage [4–6]. The increasing awareness of the importance of these emissions for earth system modeling has resulted in numerical models of regional air quality and global climate that now routinely include BVOC emissions that are estimated as a function of landcover and environmental driving variables. This is a considerable challenge due to both the hundreds of different BVOC chemical species emitted into the atmosphere [7, 8] and the vast differences in the capacity of various plant species to produce and emit terpenoids and other BVOCs [9, 10]. Furthermore, an individual compound can be emitted by different ecosystem sources that are controlled by a variety of processes. Some compounds are stored in plant tissues that are isolated from the atmosphere and are emitted only if these tissues are damaged, while other compounds are stored in structures that are open to the atmosphere and are continuously being emitted [11]. There are additional compounds that are not stored in tissues but instead are released immediately after production which may happen only in response to stress or specific environmental conditions [12].

Quantitative attempts to account for these BVOC emissions in models must consider all of the processes that control emission variability. Among the greatest of these challenges is characterizing the enormous diversity in BVOC emission types in ecosystems across the world. This paper provides an overview of our current understanding of the chemical and biological diversity of BVOC emissions into the atmosphere. Section 2describes a compilation of observations in the scientific literature that have been used to quantify BVOC emissions in a widely used numerical model, the Model of Emissions of Gases and Aerosols from Nature (MEGAN) [5], and considers the suitability of these observations for characterizing regional to global BVOC emissions. The known chemical diversity of BVOC emissions is summarized in Section 3, and an approach for improving the representation of this diversity in numerical models is described. BVOC biological diversity is discussed in Section 4, and a framework for better representation of BVOC emission diversity types is presented. Section 5 presents the major conclusions of this summary of our current understanding of the chemical and biological diversity of BVOC emissions.

BVOC EMISSION OBSERVATIONS AND MODELS

Quantitative estimation of global BVOC emissions into the atmosphere began with Went's [13] seminal work that extrapolated measurements of a single group of compounds, monoterpenes, from a single plant species,Artemisia tridentata, to the entire earth. Rasmussen [9] recognized the great diversity in BVOC emission capacities of different plants species and introduced an approach for classifying the biosphere into different vegetation groups in order to quantify regional emissions. He noted that at least some vegetation types had "fingerprints" that could be used to represent the emission behavior of those plant species. He combined estimates of USA areas of different forest types (e.g., Loblolly-shortleaf pine forest, oak-gum-cypress forest) with observations of their representative emission rates in order to quantify total BVOC emissions on a USA national and on a global scale. Zimmerman [14] extended this approach using more comprehensive land cover data including broad natural vegetation types (e.g., shrub and brush rangeland, deciduous forest, and mixed forest), agricultural lands (e.g., crops, pasture, and orchards), and a category for residential areas. This approach was limited by the large differences in the emission rates of plant species in landscapes that, for example, are classified as deciduous forest or mixed forest because of the highly variable emission rates of these broad categories of vegetation. Zimmerman made additional progress towards accounting for this by collapsing USA forests into four types: high isoprene (e.g., oak) deciduous forest, low isoprene (e.g., sycamore) deciduous forest, no isoprene deciduous (e.g., maple), and coniferous forest (e.g., loblolly pine). Lamb and colleagues [15] refined this approach using higher resolution (county scale) landcover data that included land area planted with the major crop species. This approach was extended to the global scale [16, 17] by assigning emission factors to ecosystem types in global gridded databases. This was straightforward for categories dominated by a few species (e.g., paddy rice and mangrove) but not for most categories (e.g., farm/city-cool, temperate mixed, and dry evergreen) which did not represent a uniform BVOC emission type. For regions with detailed plant species data, the Biogenic Emission Inventory System 2 (BEIS2) [1] was developed to apply BVOC emission factors for individual tree genera and crop types. However, these data were only available for forests in some regions, and BEIS2 used broad categories for grassland and shrubland ecosystem types.

The MEGAN version 2.1 (MEGAN2.1) [5] BVOC emission model assigns emission factors and parameters to 19 BVOC chemical compound classes for each of the 15 plant functional types (PFTs) used for the Community Land Model (CLM4) [18]. MEGAN2.1 can be run embedded in CLM4 and can also run offline using observations or variables from other models. BVOC emission rate measurements from about 300 studies were synthesized to estimate the emission factors used for MEGAN2.1 including data representative of the major global vegetation types. Measurements representing temperate landscapes are compiled in Table 1 [19–186]. Studies in tropical and boreal landscapes are summarized in Tables 2 [187–229] and 3 [230–268], respectively. Measurements characterizing BVOC emissions from agricultural crops are compiled in Table 4 [70, 269–282]. Terpenoid (e.g., isoprene, MBO, and monoterpene) emission factors were estimated for each of the 15 PFTs. For most of the other compounds, one or a few (e.g., one for woody PFTs and one for herbaceous PFTs) emission factors were used for all PFTs. Terpenoid emission factors are represented with a greater diversity in MEGAN2.1 both because of the greater actual diversity and because more observations have been reported.

Table 1: Compilation of studies used to estimate temperate vegetation BVOC emission factors for the MEGAN2.1 model [5]. Emission measurement approaches include enclosure (E), canopy micrometeorological (C), and landscape inverse modeling (L) techniques. Compounds include isoprene (Iso), monoterpenes (MT), sesquiterpenes (SQT), and other (Other). PFTs include broadleaf deciduous shrub (BDS), broadleaf evergreen shrub (BES), broadleaf deciduous tree (BDT), C3 grass (C3G), and needleleaf evergreen tree (NET)

Location	Approach	Compounds	PFTs	Reference
MI, USA	C	Iso	BDT	[19]
NC, USA	C	MT	NET	[20]
N T, Australia	L	Iso	BDS, BES	[21]
CO, USA	E	MT, SQT, and Other	BDT	[4]
Inner Mongolia, China	E	Iso, MT	C3G	[22]
CA, USA	C	MBO	NET	[23]
Various, USA	C	MBO, Other	NET	[24]
Austria	C	Other	C3G	[25]
NY, USA	E	MT	BES	[26]
Portugal	L	Iso, MT	BET	[27]
Various, Canada	L	Iso	BDT, NET	[28]
Potted plants	E	MT, SQT, and Other	BDS, BDT, BES, BET, and NET	[29]
FL, USA	E	Iso	BET	[30]
Potted plants	E	Other	NET	[31]
WI, USA	E	Iso	BDT	[32]
UK	E, C	Iso, MT	BES	[33]
Hangzhou, China	E	Iso, MT	BDT, BET, and NET	[34]
Italy	E, C	Iso, MT	BES, BET	[35]

Spain	E, C	MT, SQT	BET	[11]
CA, USA	E	Iso, MT	BDT, BET, and NET	[36]
Italy	C	Iso, MT, and Other	BDS, BES	[37]
Potted plants	E	MT	BES	[38]
CO, USA	C	Other	NET	[39]
Germany	E	MT	BDT	[40]
Mexico	E	Iso, MT	BDT, NET	[41]
MA, USA	L	Iso, Other	BDT	[42]
Potted plants	E	Iso, MT	BDS, BDT, BES, BET, C3G, and NET	[43]
CA	C	MT, SQT, and Other	BET	[44]
Potted plants	E	Other	BDT, BET, and NET	[45]
ON, Canada	C	Iso, MT	BDT	[46–48]
IL, USA	E	Other	C3G	[49]
Zhejiang, China	L	Iso, MT	BDT, NET	[50]
NC, USA	E, C	Iso	BDT	[51]
Various, USA	E	Iso, MT	BDT, BET	[52–54]
CO, USA	L	MBO	NET	[55]
MA, USA	C	Other	BDT	[56]
MA, USA	C	Iso	BDT	[57]
Various, USA	L	Iso, MT	BDT, NET	[58]
CO, USA	E	MT, Other	C3G	[6]
Potted plants	E	Iso, MT	BDT, BET	[59, 60]
NC, USA	C	Iso	BDT	[61]
Various, USA	E, L	Iso, MT	BDS, BDT, BES, and NET	[62–64]

TX, USA	E, C, and L	Iso, MT	BDS, BDT, and BES	[65]
Various, USA	E	Iso, MT	BDT	[66, 67]
CA, USA	E	MBO	NET	[68]
South Africa	E, C	Iso, MT	BDS, BDT, and BES	[69]
Potted plants	E	Other	BDT, C3G, and NET	[70]
Greece	E, L	Iso, MT	NET	[71]
Potted plants	E	MT	BET	[72]
Potted plants	E	Other	NET	[73]
Various, USA	E	MT, SQT	BDT, BES, BET, and NET	[74]
Various, USA	E	MT, SQT	NET	[75, 76]
ID, USA	L	MT	NET	[77]
CA, USA	C	MT	NET	[78]
Germany	E	MT, SQT	NET	[79]
Shenzhen, China	E	Iso, MT	BDS, BDT, BES, BET, C3G, and NET	[80]
WI, USA	E, C, and L	Iso, MT	BDT, C3G, and NET	[81]
Russia	E	Iso, MT, and Other	NDT, NET	[82]
Various, USA	E, C	Other	BDT, NET	[83, 84]
AZ, USA	E, L	Iso, MT, and Other	BES	[85]
CO, USA	C	Iso, MT, Other, and MBO	NET	[86]
MI, USA	C	Iso, MT, and Other	BDT, NET	[87]

CA, USA	C	MT, SQT, and Other	BDT	[88]
CA, USA	E	Iso, MT	BDS, BDT, BES, BET, C3G, and NET	[89]
Italy	E	Iso, MT, and Other	BET	[90, 91]
France	E	Iso, MT	BDT, BET	[92]
Potted plants	E	Other	BDT	[93]
FL, USA	E	MT	NET	[94]
Republic of Korea	E	MT	NET	[95]
MI, USA	C	SQT	BDT	[96]
CO, USA	E, L	MT, SQT, and Other	NET	[97]
MI, USA	E	Iso, MT, SQT, and Other	BDT, NET	[98]
Republic of Korea	L	Iso, MT, and Other	BDT, NET	[99]
VIC, Australia	E	Other	C3G	[100]
Various, China	E	Iso, MT	BDS, BDT, BES, BET, C3G, and NET	[101]
Potted plants	E	Other	BDT, C3G	[102]
Potted plants	E	Other	BDT	[103]
Switzerland	E	Other	BET, BDT	[104]
Various, USA	E, C	Iso, MT	BDT, NET	[105]
WA, USA	E, C	Iso	BDT	[106]
OR, USA	E	MT	NET	[107]
Potted plants	E	MT	NET	[108]
Shenyang, China	E	Iso, MT	NET	[109]
Republic of Korea	E	Iso	BDT, BET	[110]

Potted plants	E	Iso	BDT	[111]
Spain	E	MT	BET	[112]
FL, USA	C	Iso, MT	NET	[113]
Potted plants	E	Iso	BDT	[114]
Various, USA	E	Other	BDS, BDT, and NET	[115, 116]
MI, USA	E	MT	C3G	[117]
PA, USA	L	Iso	BDT	[118]
NM, USA	E	Iso, MT, and Other	BDT, BET, NET	[119]
Various, USA	C	Other	BDS, BDT, and BES	[120]
US, Japan, and Australia	E	SQT	BDS, BDT, BES, and BET	[121]
Japan	E	Iso, Other	NET	[122, 123]
MA, USA	C	Iso, MT, and Other	BDT	[124]
Potted plants	E	Iso	BDT	[125]
CO, USA	E	Iso	BDT	[126]
France	E	Iso	BDT	[127]
Estonia	E	Iso	BDT	[128, 129]
Portugal	E	Iso, MT	BET	[130]
Japan	E	Iso	BET	[131]
Spain	E	MT, SQT	BES, BET, and NET	[132]
MI, USA	E	Iso, MT, and SQT	BDT, NET	[133]
Various, USA	E	Iso, MT, and SQT	BDT, BET, and NET	[134]
Zambia, Botswana	E	Iso, MT	BDS, BDT, BES, and BET	[135]

Italy, France, Spain	E	Iso, MT	BDS, BDT, BES, BET, and NET	[136, 137]
NV, USA	E	Iso, MT, and SQT	BDS, BDT, BES, BET, and NET	[138]
Potted plants	E	Iso	BET	[139]
Spain	E	Iso, MT	BET	[140]
Potted plants	E	Iso	BDT	[141]
AL, USA	E	Iso	BDT	[142, 143]
Portugal	E	MT	NET	[144]
Belgium	E	Iso, MT, and Other	BDT, NET	[145, 146]
WA, USA	E	MT	NET	[147]
MI, USA	C	Iso	BDT	[148]
Austria	L	Other	C3G	[149]
Italy	E	Iso, MT	BDT, BET	[150]
Various, USA	E	Iso	BDT, BET, and NET	[151]
Various, USA	E, L	Iso, MT	BDT, BET, C3G, and NET	[152]
CA, USA	E	Other	C3G	[153]
CO, USA	L	MT	BDT, NET	[154, 155]
Potted plants	E	Iso, MT	BDT	[156]
Potted plants	E	Other	NET	[157]
Potted plants	E	Other	BDT, NET, and C3G	[158]
CA, USA	C	MBO, Other	NET	[159]
Potted plants	E	MT, Other	BET	[160]
NC, USA	E	Iso	BDT	[161]
Various, USA	E	Iso	BDT	[162]
Nepal	L	Iso, MT	NET, BDT, and C3G	[163]

Georgia, USSR	E, L	Iso, MT	BDT, NET	[164]
Belgium	E	MT, Other	BDT	[165]
Various, USA	L	Other	BDT, NET	[166]
Germany	C	Iso, MT, and Other	BDT	[167]
Potted plants	E	MT	BET	[168]
Germany	C	Iso, MT	BDT, NET	[169]
Japan	E	Iso	BDT, BET	[170]
Japan	C	MT	NET	[171]
Potted plants	E	Iso	BET	[172]
Potted plants	E	MT	NET	[173]
Greenhouse	E	Iso, MT	BET, BDT	[174]
Inner Mongolia, China	E	Iso, MT	C3G	[175]
Various, USA	L	Iso, MT	BDT, NET	[176]
France	E	Iso, MT	BDT, BET	[177]
Potted plants	E	MT	BES	[13]
MI, USA	C	Iso	BDT	[178]
TX, USA	L	Iso, MT	NET, BDT	[179]
Various, USA	L	Iso, MT	BDT, C3G	[180]
WA, Australia	E	Iso, MT, and Other	BET	[181]
Beijing, China	E	Iso, MT	BDT	[182]
Potted plants	E	Other	NET	[183]
Japan	L	MT	NET	[184]
Potted plants	E	MT	NET	[185, 186]
Various, USA	E	Iso, MT, and Other	BDS, BDT, BES, BET, C3G, and NET	[14]

Table 2: Compilation of studies used to estimate tropical vegetation BVOC emission factors for the MEGAN2.1 model [5]. Emission measurement approaches include enclosure (E), canopy micrometeorological (C), and landscape inverse modeling (L) techniques. Compounds include isoprene (Iso), monoterpenes (MT), and other (Other). PFTs include broadleaf deciduous tree (BDT), broadleaf evergreen tree (BET), and warm C4 grass (C4G)

Location	Scale	Compounds	PFTs	Reference
Yunnan, China	C	Iso, MT	BDT	[187]
Malaysia	E, L	Iso, MT	BDT, BET	[188]
AM, Brazil	L	Iso, MT	BDT, BET	[189]
Venezuela	L	Iso	C4G	[190]
Costa Rica	E, C	Iso, MT, and Other	BDT, BET	[191]
AM, Brazil	C	Iso, MT	BDT, BET	[192]
CAR	L	Iso	BDT, C4G	[193]
Botswana	C	MT	BDT	[194]
Various, Brazil	L	Iso, MT	BDT, BET	[195]
Guyana	L	Iso	BDT, BET	[196]
South Africa	E	Iso, MT	BDT, BET	[197]
Various, Brazil	E	Iso, MT	BDT, BET	[198]
Peru	L	Iso, MT	BDT, BET	[199]
Venezuela	L	Iso, MT, and Other	C4G	[200]
Costa Rica	C	Iso, MT, and Other	BDT, BET	[201]
AM, Brazil	C	Iso, MT, and Other	BDT, BET	[202]
AM, Brazil	C	Iso, MT, and Other	BDT, BET	[203]
Panama	E	Iso	BDT, BET	[204]
AM, Brazil	L	Iso, MT	BDT, BET	[205]
RO, Brazil	E, C	Iso, MT, and Other	BDT, BET	[206]
Cameroon, CAR, and Congo	E	Iso, MT	BDT, BET	[207]

Yunnan, China	E	Iso, MT	BDT, BET	[101]
RO, Brazil	E	Iso, MT	BDT	[208]
AM, Brazil	E	Iso, MT	BDT, BET	[209]
AM, Brazil	C	Iso, MT	BDT, BET	[210]
Sabah, Malaysia	C	Iso, MT, and Other	BDT, BET	[211, 212]
Potted plants	E	Iso, MT	BDT, BET	[213]
Sabah, Malaysia	C	Iso, MT, and Other	BET	[214, 215]
PA, Brazil	C	Iso, MT	BDT, BET	[216]
Potted plants	E	Iso	BDT, BET	[217]
India	E	Iso	BDT, BET	[218]
Panama	E, L	Iso	BDT, BET	[151]
AM, Brazil	L	Iso	BDT, BET	[219]
PA, Brazil	C	Iso, MT	BDT, BET	[220]
Malaysia	E, L	Iso, Other	BDT, BET	[221]
Venezuela	L	Iso	C4G	[222]
Benin	E, L	Iso, MT	BDT, BET	[223]
Congo	C	Iso	BDT, BET	[224]
India	E	Iso	BDT, BET	[10]
AM, Brazil	E, L	Other	BDT, BET	[225]
India	E	Iso	BDT, BET	[226]
Surinam	L	Iso, MT, and Other	BDT, BET	[227, 228]
Nigeria, AM, Brazil	L	Iso, MT	BDT, BET	[229]

Table 3: Compilation of studies used to estimate boreal vegetation BVOC emission factors for the MEGAN2.1 model [5]. Emission measurement approaches include enclosure (E), canopy micrometeorological (C), and landscape inverse modeling (L) techniques. Compounds include isoprene (Iso), monoterpenes (MT), sesquiterpenes (SQT), and other (Other). PFTs include broadleaf deciduous shrub (BDS), broadleaf deciduous tree (BDT), arctic C3 grass (AC3), needleleaf deciduous tree (NDT), and needleleaf evergreen tree (NET)

Location	Scale	Compounds	PFTs	Reference
Finland	E	MT	NET	[230]
Jilin, China	C	Iso, MT	BDT, NET	[231]
Sweden	E	Iso	AC3	[232, 233]
Sweden	E	Iso	AC3	[234]
Potted plant	E	Iso, MT, and Other	NET	[235]
SK, Canada	C	Iso	BDT	[236]
Finland	C	MT	BDT, NDT, and NET	[237]
Finland	C	Iso, Other	AC3	[238]
Finland	E	MT, SQT	BDS, BDT	[239]
Finland	E, L	Iso, MT, and SQT	BDT	[240, 241]
Finland	E	MT, SQT	NET	[242]
Potted plants	E	Iso	AC3	[243]
Finland	E	Iso, MT, and Other	AC3	[244]
Sweden	C	Iso, MT, and Other	AC3	[245]
Norway	L	MT	NET	[246]
WI, USA	E, C, L	Iso, MT	NDT	[81]
Various, Russia	E	Iso, MT, and Other	BDT, NDT, and NET	[82]

Sweden	L	MT	NET	[247]
Sweden	E	Iso	AC3, NET	[248]
Sweden	E	Iso, MT, and Other	AC3, NET	[249]
ON, Canada	L	Iso	BDT, NET	[250]
Potted plants	E	Iso, MT	NET	[251]
ON, Canada	E	Iso, MT	AC3	[252]
Jilin, China	E	Iso, MT	AC3, BDS, BDT, NDT, and NET	[101]
Potted plants	E	MT, SQT	NET	[253]
SK, Canada	C	Iso	BET	[254]
Sweden	E, L	MT	NET	[255]
AL, USA	C	Iso, MT, and Other	AC3, BDS	[256]
Finland	E, C	MT	NET	[257]
Finland	C	MT	NET	[258]
Finland	E, C	Iso, MT	BDT, NET	[259]
Finland	C	Iso, MT, and Other	NET	[260]
Finland	E	Iso, MT, and SQT	NDT	[261]
Finland	L	Iso, MT	BDT, NET	[262]
Finland	E	MT, SQT	NET	[263]
Sweden	E	Iso	AC3	[264, 265]
Finland	E	MT, SQT, and Other	BDS, BDT	[266]
Sweden	E	MT, SQT, and Other	BDS, BDT	[267]
Various, Canada	C	Iso	AC3, BDS, BDT, and NET	[268]

Table 4: Compilation of studies used to estimate cropland BVOC emission factors for the MEGAN2.1 model [5]. Emission measurement approaches include enclosure (E), canopy micrometeorological (C), and landscape inverse modeling (L) techniques. Compounds include isoprene (Iso), monoterpenes (MT), sesquiterpenes (SQT), and other (Other)

Location	Scale	Compounds	Crop	Reference
Potted plants	E	MT, SQT, and Other	Potato	[269]
CA, USA	E	Iso, MT, and Other	Various	[270, 271]
CA, USA	E	Iso, MT, and Other	Various	[272]
UK	E	Iso, MT, and Other	Miscanthus, willow coppice	[273]
Potted plants	E	SQT, Other	Tobacco	[274]
Potted plants	E	Iso, MT, SQT, and Other	Switchgrass	[275]
Potted plants	E	SQT	Corn	[276]
Potted plants	E	Other	Sorghum	[70]
Potted plants	E	Iso	Arundo donax	[277]
Spain, Italy	E	Other	Corn, pea, barley, and oat	[93]
VIC, Australia	E	Other	Clover	[100]
Potted plants	E	Iso, MT, SQT, and Other	Wheat, rye, rape, and grape	[102]
Potted plants	E	Iso	Velvet bean, kudzu	[278]
Potted plants	E	Other	Soybean, tomato, bean, and Corn	[115]

Potted plants	E	SQT	Corn	[279]
Potted plants	E	SQT, Other	Sunflower	[280]
Italy	E	Other	Fescue	[281]
CO, USA	C	Other	Alfalfa	[282]
Various, USA	E	Iso, MT, Other	Various	[14]

Until recently, most BVOC emission measurements were conducted using enclosure techniques, but whole canopy flux measurements using micrometeorological approaches are now becoming more common [215]. Characterizing BVOC emissions with enclosure measurements is challenging due to difficulties in accessing all parts of a mature forest canopy and because of the presence of storage structures which can be disturbed resulting in emissions at rates much higher than for undisturbed conditions [283]. These issues resulted in BVOC emission factors reported by earlier studies that greatly underestimate isoprene emissions, because isoprene emission rates are lower for the shaded leaves in the more easily accessed portion of a forest canopy, and overestimate monoterpene emissions because of disturbances to terpenoid storage structures [284]. The above-canopy flux measurements integrate over the entire canopy and landscape without disturbing emission rates [86]. Capabilities for quantifying biogenic VOC fluxes have steadily improved over the past decades including recent analytical advances such as the time-of-flight proton-transfer reaction mass spectrometer (PTR-TOF-MS) that enables whole canopy measurement of a wide range of BVOC fluxes [285]. Aircraft VOC flux systems have footprints of several km and can characterize fluxes over entire domains of hundreds of km and so are suitable for evaluating fluxes estimated by regional models [286]. Tower-based VOC flux systems typically have a footprint of hundreds of meters and are well suited for quantifying diurnal, seasonal, and interannual variations. Biogenic VOC fluxes have been measured at more than 45 tower locations (Tables 1 to4 and summarized in [287, 288]), but most of these studies were for a short period (a few weeks or less) of time. The availability of more than 500 above-canopy flux towers constructed for water, carbon, and energy flux studies provides an opportunity to add biogenic VOC measurements without the cost of basic site development [289]. Measurements at a large number of sites can be accomplished with low-cost and low-power relaxed eddy accumulation measurements systems [238].

Figure 1 shows that there were relatively few BVOC emission rate observations reported in the 1960s and 1970s, and all but one of these studies were in temperate regions. Interest in the role of BVOC emissions in regional ozone pollution in the 1970s [9] stimulated publications on this topic by the early 1980s including some investigations of tropical, boreal, and agricultural ecosystems. This interest peaked in the mid-1980s and then declined as some researchers concluded that BVOC emissions did not have an important role in regional air quality [290]. An improved understanding of the magnitude of BVOC emissions and the relatively high sensitivity of ozone to BVOC emissions demonstrated that this was not the case [291, 292] and led to a resurgence in BVOC emissions research in boreal, tropical, and agricultural ecosystems in the 1990s. Interest in tropical landscapes was driven by the recognition that the tropics are responsible for 80% of global emissions [17]. There was initially little interest in BVOC from agricultural ecosystems because of the generally low terpenoid emissions from these plant species, but the discovery of substantial amounts of oxygenated VOC emissions from crops [102, 115] led to more studies. The annual publication rate decreased in the mid-2000s, but there has been a recent increase in the number of publications. This has likely been driven by the recognition of the important role of BVOC in secondary organic aerosol production [3, 293].

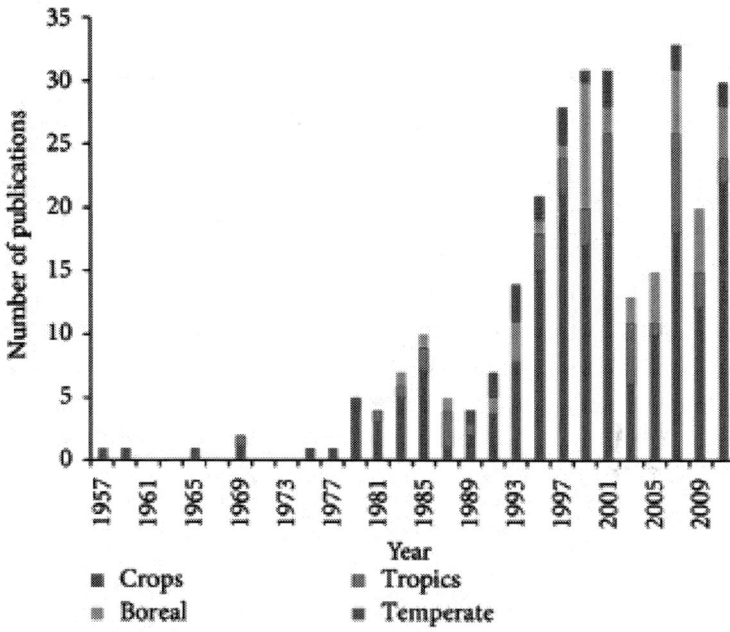

Figure 1: Illustration of the number of publications during two-year periods that were used to develop the BVOC emission factors for the MEGAN2.1 model [5].

Figure 2 shows that temperate, tropical, and boreal ecosystems each cover 25 to 35% of the global vegetation-covered land surface with croplands covering the remaining 15%. This figure also shows that although the estimated global BVOC emission is dominated by tropical ecosystems, most studies have focused on temperate ecosystems. Needleleaf trees, broadleaf trees, shrubs, grass and crops each cover 10 to 30% of the global vegetation-covered land area but broadleaf trees are estimated to contribute nearly 80% of the emissions (Figure 3). Investigations of BVOC emission have generally focused on the important emission sources although broadleaf trees are somewhat understudied. Figure 2 shows that isoprene contributes about half of total emissions and was also investigated in about half of these studies. In contrast, other VOC are 36% of the estimated emissions and were examined in only ~20% of the studies. Recent studies provide some balance with relatively more investigations in the tropics and measurements of other VOC (Figure 1).

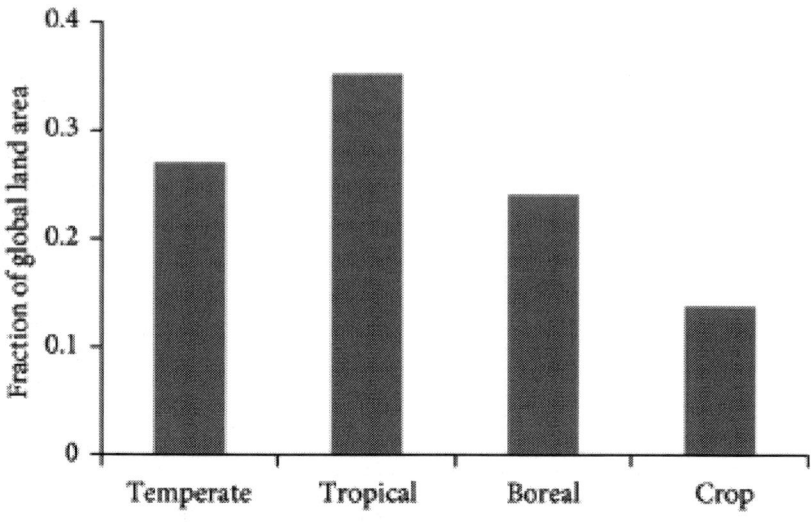

(a)

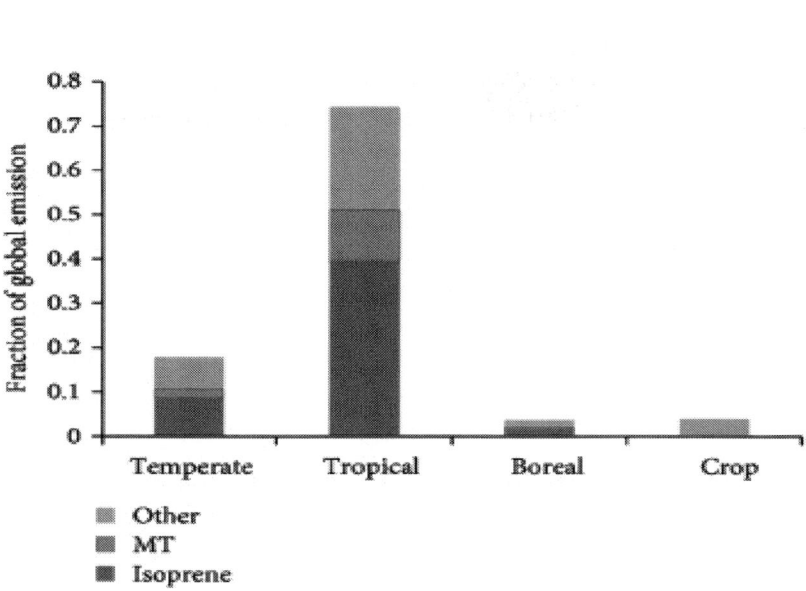

(b)

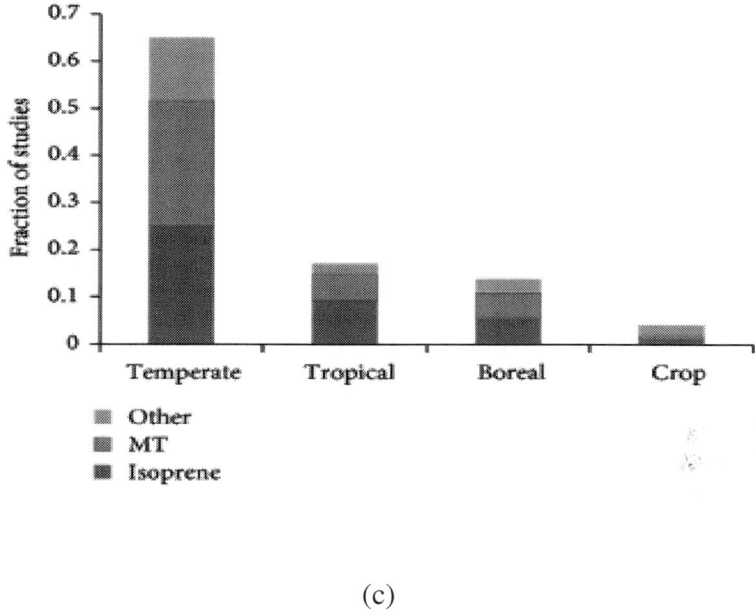

(c)

Figure 2: Comparison of the global fraction of vegetation-covered land area, global BVOC emissions estimated by [5], and the number of BVOC emission diversity studies (Tables 1–3) for major biomes types (temperate, tropical, boreal, and crop) and compound categories (isoprene, monoterpene, and other).

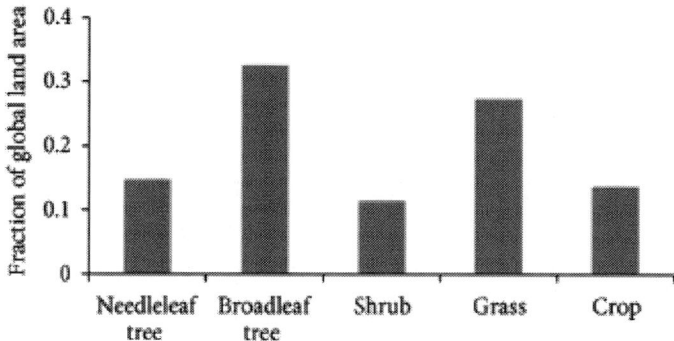

(a)

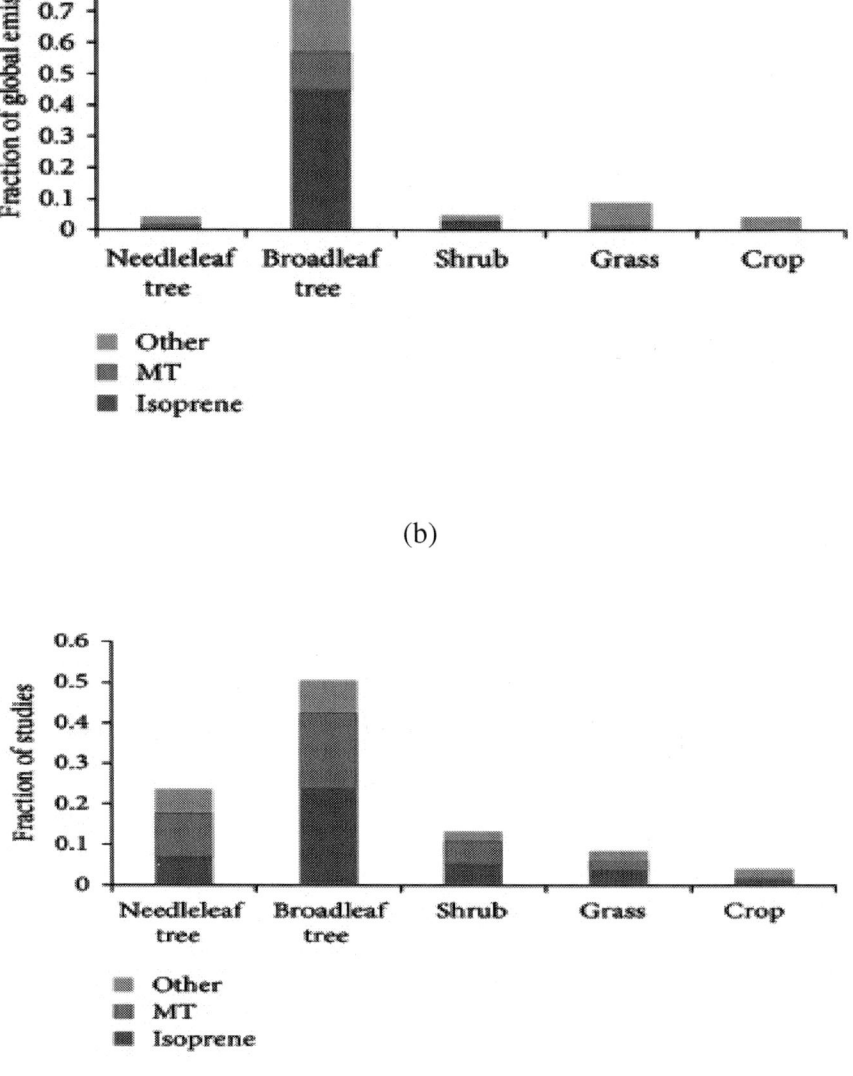

(b)

(c)

Figure 3: Comparison of the global fraction of vegetation-covered land area, global BVOC emissions estimated by [5], and the number of BVOC emission diversity studies (Tables 1–3) for major plant functional types (needleleaf tree, broadleaf tree, shrub, grass, and crop) and compound categories (isoprene, monoterpene, and other).

BVOC CHEMICAL DIVERSITY

Terrestrial ecosystems produce thousands of chemical species that can be emitted into the atmosphere [294] but only a few of these compounds are emitted at the rates required to have a significant impact on atmospheric composition [5]. Most of these chemicals are organic compounds including some that contain oxygen, nitrogen, sulfur, or halogens. Biogenic emission models often reflect this dominance by including only a few major compounds, such as isoprene and α-pinene, and omitting the rest or including them as a generic undefined "other" category. More recently, the MEGAN2.1 biogenic emission model [5] was developed to estimate emissions of 147 compounds that were thought to be significant or potentially significant. This section describes BVOC chemical diversity and potential improvements over the MEGAN2.1 scheme.

Terpenoid Compounds

Terpenoid compounds have long been considered the dominant global BVOC [9]. This incredibly diverse group includes thousands of chemical species that can be classified as hemiterpenoids (C5), monoterpenoids (C10), sesquiterpenoides (C15), homoterpenes (C11 and C16), diterpenoids (C20), and larger compounds with such low volatility that it is unlikely that they are emitted into the atmosphere in a gaseous form. Terpenoids include oxygenated terpenes such as the hemiterpenoid methylbutenol (MBO), the monoterpenoid linalool, and the sesquiterpenoid cedrol. These oxygenated terpenoids are a small portion of the global total terpenoid emission but may be important in some regions. About half of the 147 BVOC species included in MEGAN2.1 are terpenoid compounds including some that are major contributors to global BVOC emissions (e.g., isoprene, α-pinene) and others that are minor components.

Investigations of BVOC began centuries ago with interest in commercial applications for monoterpenes in the flavor and fragrance industries. These activities led to the development of diverse analytical techniques and a considerable body of the literature describing terpenoid production and distribution in the oleoresins stored within plant tissues. Very little of this information has been incorporated into BVOC emission models because the production of monoterpenes by plants and their release into the atmosphere are not always well correlated, and only a small fraction of the hundreds of monoterpene compounds identified in essential oils have been observed as significant atmospheric BVOC emissions. Some monoterpenoids are oxygenated compounds including some multifunctional oxygenates and acetylated compounds that may make a disproportionate contribution to secondary aerosol production. Early studies indicated that a few monoterpenes (α-pinene, β-pinene, limonene, sabinene, 3-carene, and myrcene) dominated the total monoterpene flux into the atmosphere [284]. However, these studies typically did not attempt to measure all compounds, and some monoterpenes may have been reported more frequently simply because these were the only compounds targeted. It was initially thought that all monoterpenes emanated from storage pools and were controlled only by leaf temperature. The discovery of high emission rates of light-dependent monoterpene emissions, produced from recently synthesized carbon in a manner similar to isoprene, from European [90] and African [194] savannas, tropical forests [202], and boreal needleleaf trees [237] led to the introduction of multiple emission processes for an individual chemical species in BVOC emission models.

Organic chemists investigating monoterpenes in the late 1800s identified the hemiterpene, isoprene (2-methyl-1,3-butadiene), as the biochemical precursor of monoterpenes, but isoprene was thought to exist only within plant tissues [151]. The discovery of substantial isoprene emissions from plants into the atmosphere was discovered more than 50 years ago and was initially controversial [152]. Isoprene later became recognized as the dominant global BVOC emission into the atmosphere [17]. Isoprene contributes about half of the total global BVOC flux, and so it is not surprising that it has been investigated more extensively than any other atmospheric BVOC.

Sesquiterpenes (SQTs) are a major component of essential oils stored by some plants, especially broadleaf trees, and can also be directly emitted without being stored [295]. SQTs are emitted from numerous plant species including conifer and broadleaf trees, shrubs, and agricultural crops [296]. While some sesquiterpenes, such as longifolene, have atmospheric oxidation lifetimes on the order of hours, similar to that of the dominant monoterpenes such as α-pinene, some of the most dominant sesquiterpenes emitted into the atmosphere (β-caryophyllene and farnesene) are much more reactive and have typical lifetimes of minutes [297]. The low volatility, and in some cases high reactivity, of sesquiterpenes makes them considerably more difficult to detect and quantify. As a result, few studies considered sesquiterpene emission measurements since they were generally thought to be a minor contribution in comparison to monoterpenes. Efforts to quantify sesquiterpene emissions increased in the past decade with the growing interest in atmospheric secondary organic aerosol [76]. Although sesquiterpenes are only a minor fraction of total BVOCs, they are recognized as important for the atmosphere due to their relatively high SOA yields [298].

Large emissions of an oxygenated hemiterpene, 2-methyl-3-buten-2-ol (referred to here as MBO) were observed from pine trees in the early 1990s although emissions of MBO from insects and flowers had been observed previously [55]. MBO is emitted at high rates from some pine species, such as Pinus ponderosa, and low rates from other pines, including most Eurasian pines [299]. The global MBO emission is less than 1% of the global total BVOC, but MBO is the dominant emission in ecosystems dominated by high MBO emitting pines including large areas of western USA forests. Recent studies suggest that MBO may be emitted from most isoprene emitting vegetation at a rate that is ~1% of the isoprene emission rate [300]. This low level emission over a large part of global terrestrial ecosystems could be of the same magnitude as the localized emission from high MBO emitters.

The production of some terpenoid compounds is elevated in response to stress and is often observed as a light dependent, de-novo emission [301]. These include monoterpenes (e.g., ocimene), oxygenated monoterpenes (e.g., linalool), sesquiterpenes (e.g., farnesene), the homoterpenes dimethyl-nonatriene (DMNT), and trimethyl-tridecatetraene (TMTT). Emissions of these compounds are not always present, but when they are observed they can exceed typical monoterpene or sesquiterpene emission rates. The large variability and limited knowledge of factors controlling stress-induced BVOC emissions result in high uncertainties associated with emissions of these compounds, but they may be a substantial component of total BVOC emissions into the atmosphere, and a better understanding is needed.

Methanol and Acetone

Methanol and acetone are among the most abundant VOCs in the global atmosphere. High concentrations of atmospheric methanol and acetone observed by investigators in the 1960s were attributed primarily to the atmospheric oxidation of VOC with minor contributions from bacteria, biomass burning, and anthropogenic sources [302]. In the early 1990s, high rates of methanol emissions were observed from vegetation foliage, especially young expanding leaves [115]. Lower rates of acetone emissions were observed from conifer buds [116]. A few years later, decaying leaf litter was found to be a smaller but significant abiotic source of methanol and acetone [303].

Jacob et al. [304] estimated that terrestrial ecosystems (biotic and abiotic) dominate global methanol emissions with 78% of the global annual production with the remainder being from atmospheric oxidation of VOC (15%), biomass burning (5%), and urban (2%) sources. Millet et al. [305] used additional in situ observations and concluded that oceans were responsible for 35% of the global methanol emission and assigned a contribution of 42% to terrestrial ecosystems. Stavrakou et al. [306] used both satellite and aircraft observations to constrain global methanol distributions and report annual emissions of 187 Tg per year with a contribution of 53% from vegetation. They also identified missing sources in arid and semiarid regions of Central Asia and Western USA.

An analysis of the global acetone budget by Jacob et al. [307] indicated contributions to total emissions from terrestrial ecosystems (37%), atmospheric oxidation of VOC (29%), ocean (28%), biomass burning (5%), and anthropogenic emissions (1%). A more recent analysis concluded that terrestrial ecosystems were responsible for only 22% and oceans contributed 55% [308].

Acetaldehyde, Formaldehyde, Ethanol, and Organic Acids

Kesselmeier [309] described both the atmospheric importance of short-chained oxygenated VOCs (e.g., acetaldehyde, formaldehyde, acetic acid, and formic acid) and the challenge of quantifying their atmospheric budgets. This includes the following challenges: (1) there are both natural and anthropogenic sources of these compounds, (2) there are primary and secondary (atmospheric oxidation) sources, (3) these compounds are difficult to measure, and (4) vegetation is both a source and a sink of these compounds. The strong bidirectional exchange exhibited by these compounds requires that both emission and deposition need to be considered. Accurate simulation of land-atmosphere fluxes of these compounds requires estimates of their atmospheric concentrations and the compensation point for each compound.

Alcoholic fermentation in the leaves and roots of plants produces ethanol which is converted to acetaldehyde in a pathway leading to acetate consumption [83]. Millet et al. [310] identified the major sources of atmospheric acetaldehyde as oxidation of VOC (60%), ocean (27%), and terrestrial ecosystems (11%). Biomass burning and anthropogenic emissions contribute the remaining 2%. The introduction of the PTRMS technique has provided an increasing number of measurements of acetaldehyde emissions from vegetation, including whole canopy flux measurements, while there remain relatively few data for ethanol [310].

Substantial emissions of formaldehyde, and lesser amounts of formic and acetic acid, have been reported from studies using enclosure measurements to investigate various tree species [103, 309]. While emissions can be considerable, there is also the potential for a strong uptake of these compounds. These enclosure measurements suggest that the net flux of these compounds is a small emission into the atmosphere. Recent studies using above-canopy measurements have provided evidence that formaldehyde and formic acid emissions could be much larger. An analysis of satellite data suggests that formic acid emissions are two to three times higher than estimated from known sources [311] and that 90% of formic acid has a biogenic origin which includes direct emission and production from terpenoids. The first whole canopy fluxes of formaldehyde measured by eddy covariance have recently been reported [39]. The above-canopy flux, a net emission, was much higher than predicted from enclosure measurements which may be because the flux included both primary emissions and within canopy production. Measurements to better constrain formic acid and formaldehyde fluxes are needed.

Stress Compounds

Environmental and biotic stresses are important factors controlling BVOC emissions [128]. This includes BVOCs that are emitted at relatively low levels with unstressed conditions and then are elevated under stressed conditions (e.g., α-pinene) and compounds that are typically observed only when plants are stressed (e.g., methyl salicylate). BVOCs associated with pathogen or herbivore-induced stress include ethene, methanol, terpenoids, benzenoids, and green leaf volatiles [73, 128, 312–315]. The biochemical pathways and the defensive roles of these compounds have been the subject of many investigations [316], but there have been few attempts to quantify these emissions and they have not been integrated into regional BVOC emission models. The current limited understanding of the processes controlling stress-induced emissions makes any numerical approach for estimating stress BVOC emissions highly uncertain. Observations that provide an initial assessment of stress-induced emissions provide a first step towards assessing their contribution to total BVOC emissions and the need for accounting for these processes in BVOC emission models.

Ethene is an important phytohormone, and its emission rate from plants has been used as an indicator of stress [317]. Sawada and Totsuka [158] estimated an annual global flux of 18 to 45 Tg of ethene with 74% released from natural sources. This was the first global emission estimate of a nonterpenoid BVOC and was based on an extrapolation of enclosure measurements that indicated widespread ethene production by plants in most landscapes. Canopy scale fluxes measured above a temperate deciduous forest confirmed that substantial amounts of ethene were released into the atmosphere from this landscape [56]. The canopy scale fluxes are in reasonable agreement with the earlier enclosure measurements.

The green leaf volatiles are a major category of BVOC that is associated with plant response to herbivory and other stresses [12]. These compounds are produced in plants from linoleic and linolenic acid which are unsaturated fatty acids. The most prominent of these with respect to emissions into the atmosphere are cis-3-hexenal, trans-2-hexenal, hexanal, 1-hexanol, and cis-3-hexenol [287]. The compound methyl jasmonate is also produced from this pathway and has an important role in plant signaling [318].

Leaf Surface Compounds

Leaf surfaces are covered by a waxy material that serves as a barrier for keeping water in and pathogens out [319]. Long-chain hydrocarbons, acids, alcohols, and esters are the dominant components of these leaf waxes, but there are a variety of other constituents [320]. While these high molecular weight compounds have low volatility, a small fraction can volatilize into the gas phase, and this may be significant, especially with the high leaf temperatures (>40°C) that occur in hot deserts. A study by Matsunaga et al. [120] concluded that some compounds, including homosalate and 2-ethylhexyl salicylate, were emitted at significant rates from a wide variety of plants. These are sunscreen compounds that protect plant tissues from UV solar radiation [120]. The estimated contribution to total emissions from most ecosystems was small, but a large contribution was estimated for desert regions dominated by mesquite (Prosopis spp.) which is an important component of large areas in the southwestern USA.

Another source of VOC emissions from vegetation is the oxidation of organics on the surface of leaves and other structures. Fruekilde et al. [45] fumigated leaves with ozone and reported elevated emissions of 6-methyl-5-hepten-2-one, acetone, geranyl acetone, and 4-oxopentanal and suggested that ozonolysis at vegetation surfaces could explain the widespread occurrence of these compounds in ambient air. Karl et al. [321] noted that elevated oxygenated VOC emission from foliage exposed to ozone could also be due to increased production of these compounds in leaves in response to stress or to gas phase oxidation (secondary compounds). They conducted experiments to isolate the mechanisms responsible for oxygenated VOC production and concluded that a substantial amount of oxygenated VOC was primary emissions, originating from reaction of ozone inside of the plant or on plant surfaces, although there were also some secondary products from gas phase reactions.

Organic Halides

Organic halides including methyl bromide, methyl chloride, and methyl iodide are produced by vegetation and emitted into the atmosphere. Emissions are controlled by environmental conditions including soil moisture and temperature [322]. Even though methyl halide fluxes are small compared to terpenoid emissions, they are an important source of halogens in the stratosphere where they play a role in stratospheric ozone depletion [153]. Quantifying fluxes of methyl halides is challenging because terrestrial ecosystems are both a source and a sink of these compounds [166, 323]. Stable isotopes are now being used to individually quantify gross emission and uptake rates to improve understanding of the processes driving net fluxes [322, 324].

Organic Sulfur Compounds

Biogenic organic sulfur emissions from marine and terrestrial ecosystems are an important source of atmospheric sulfur compounds in clean environments [325]. Soil microbes and plants are both sources of compounds that include methyl mercaptan, dimethyl sulfide, and dimethyl disulfide. A more recent study [326] estimated that terrestrial ecosystems contribute about 15% of the global dimethyl sulfide flux with the remainder coming from oceans. Higher weight organic sulfur compounds such as diallyl disulfide, methyl propenyl disulfide, and propenylpropyldisulfide can be emitted in substantial amounts from a few plant species [149].

Alkanes (including Oxygenated Alkanes)

Zimmerman [14] reported that a variety of alkanes were a substantial fraction of the biogenic VOCs emitted from vegetation. This was based on gas chromatograph retention times, rather than identification by mass spectrometry, and later studies have found only very low level of emissions of alkanes including ethane [100], propane [249], pentane [82], hexane [136], heptane [157], C6 to C10 saturated aldehydes [327], alcohols [281], ketones [282], pyruvic acid [85], and methane [328]. The potentially large source of methane [328] has been controversial as following studies found either much lower or no methane emission from living plants [329]. Terrestrial ecosystems are, however, a major source of methane emission from soil microbes and termites [328].

Benzenoid Compounds

The extensive BVOCs emission surveys of Zimmerman [14] also indicated that benzenoid compounds were a substantial fraction of total BVOC emissions. As was the case for alkanes, later studies found much lower benzenoid emissions. However, it is widely recognized that there are many benzenoid compounds (e.g., benzaldehyde, anisole, and benzyl alcohol) emitted as floral scents [330]. These floral benzenoid emissions are thought to make a small contribution to annual regional BVOC emissions [4] but can be a major emission at specific locations [214]. At least some of these compounds (e.g., toluene and methyl salicylate) are associated with plant stress and have been observed at elevated rates from stressed plants [73, 88].

Other Alkenes (including Oxygenated Alkenes)

The terpenoids are not the only alkenes emitted into the atmosphere from terrestrial ecosystems. Propene and butene emissions have been observed in enclosures and confirmed by above-canopy flux measurements [56]. Other longer-chain alkenes have only been observed using enclosure techniques. This includes 1-dodecene and 1-tetradecene [270]. Oxygenated alkenes such as 1, 3-octenol [281], neryl acetone [75], terpinyl acetate [183], and nonenal [74] have also been observed but are thought to be minor in comparison to terpenoid emissions.

Representing BVOC Chemical Diversity in Numerical Models

The first detailed biogenic VOC emission inventory [9] included estimates of just two compounds: isoprene and α-pinene. Several decades later, the USA EPA released the first widely available biogenic emission inventory approach, called BEIS [331]. In addition to emission of isoprene and α-pinene, BEIS included lumped categories for "other monoterpenes" and an "unidentified" category. While this made the emission inventory more comprehensive, the "unidentified" category had limited use in atmospheric chemistry models because BVOCs have such varied atmospheric impacts (e.g., a wide range in aerosol yields and ozone production potential). In addition, some highly reactive BVOC may control the local atmospheric oxidizing capacity, while other less reactive compounds are transported long distances to remote areas or to the stratosphere where they can impact the chemistry of these pristine regions. An initial attempt to account for this was made [17] by using two "other" BVOC categories that included "other reactive VOC," such as 232-MBO and "other VOC" which included less reactive compounds such as methanol and acetone. Emissions of 39 individual BVOCs were later estimated [287] in addition to three other categories: other terpenoids, other reactive NMVOCs, and other NMVOCs. The 39 identified compounds contributed about 94% of the total emission. MEGAN2.1 [5] eliminated the use of any "unidentified" categories and estimated emissions of 149 known compounds.

Most atmospheric chemistry schemes include at most only a few BVOCs and may lump these together with other compounds which limits the advantages of a detailed emissions chemical speciation. The increased number of compounds is a disadvantage if there is a significant increase in the computational resources associated with emissions parameterization, processing inputs, and emission calculations. MEGAN2.1 [5] uses a balanced approach that includes individual representations of 13 major BVOCs along with 5 additional categories for which an emission was calculated, and then the total was speciated into individual BVOC. This approach required the calculation of the emission activity of 18 BVOC types. The emission behavior of a compound, for example, the light dependent response, was treated the same for all vegetation types. This is reasonable for some compounds, such as isoprene, but not for others, such as α-pinene, which have different emission behavior in a tropical forest than in a temperate needleleaf forest [202]. This approach can be improved by using a smaller number of compound types but allowing a different emission behavior for different vegetation types. The 18 BVOC categories used for MEGAN2.1 [5] could be reduced to about half that number. For example, a nine BVOC category scheme could include hemiterpenes, light-dependent monoterpenes, light-independent monoterpenes, sesquiterpenes, methanol, acetone, bidirectional compounds, stress compounds, and other compounds. Each of these nine BVOC emission categories could have a different speciation profile for each vegetation type to simulate differences such as the contributions of individual monoterpenes to the total monoterpene flux from different landscapes.

BVOC BIOLOGICAL DIVERSITY

Just as the scent of various flowers can be quite distinct, the total BVOC emission rates of various plants can differ. Some plants have total BVOC emission rates that are less than 0.01 μg per gram (dry weight) per hour ($\mu g\, g^{-1}\, h^{-1}$), while others have rates that exceed $100\, \mu g\, g^{-1}\, h^{-1}$. In addition to the three orders of magnitude variability in total emission, chemical composition can vary greatly with some plants dominated by isoprene, while emissions of other plants are dominated by other compounds such as α-pinene, MBO, or methanol. The BVOC emission rates of different terrestrial ecosystems vary by more than three orders of magnitude because the landscape average emission is determined both by the variability associated with plant-specific BVOC emission rates and the variability in vegetation cover fraction. In order to investigate BVOC emission variations associated with biological diversity, it is useful to define an emission factor for a set of standard conditions such as leaf age, growth environment, light, temperature, CO_2 concentration, soil moisture, and others [5, 129]. While there are clear taxonomic patterns associated with BVOC emissions, with plants of the same species or genus tending to be more similar, there are also many exceptions [332, 333]. This is not unexpected since the taxonomic schemes used to classify plants are not based on their BVOC emissions characteristics. In addition, some BVOC emissions variability is expected within plant species. For example, pine trees emit a variety of monoterpenes that are used for chemical defense against many different pests [334]. If all individuals of a pine species emit the same mix of monoterpenes, then a herbivore that manages to overcome this particular chemical mixture could devastate that pine species. If there are pine populations with different monoterpene emission types, then at least some pine tree individuals will survive.

Welter et al. [177] investigated BVOC emissions of an isoprene emitting oak species (Quercus canariensis), a monoterpene emitting oak species (Q. suber), and a species that is a hybrid of those two oak species (Q. afares). They found that Q. afares individuals were monoterpene emitters but at relatively low rates and with high variability. Geron et al. [54] examined isoprene emissions from Populus hybrids and found that their emission factors were a factor of two higher than their parents and that the second generation crosses had even higher emission factors. Bäck et al. [230] measured terpenoid emissions of individual Scots Pine (Pinus sylvestris) trees in a forest stand in Finland. They found that emissions of some trees were dominated by α-pinene, while others emitted primarily D3-carene, and still others emitted similar amounts of both. These studies demonstrate that there can be substantial within-species variation in terpenoid emissions for at least some plant species. Geron et al. [54] also considered whether there were significant interspecies differences in the isoprene emission factors of isoprene-emitting temperate broadleaf tree species and concluded that there was no clear evidence of this. Variability within and between species was similar suggesting that all temperate broadleaf trees could be divided into just two categories with respect to their isoprene emission: low emitters ($<1\ \mu g\,g^{-1}\,h^{-1}$) and high emitters (about $90\ \mu g\,g^{-1}\,h^{-1}$). Isoprene emission factors for high emitting temperate needleleaf trees were much lower than broadleaf trees indicating a need to assign different isoprene emission factors to different PFTs.

Numerical land surface models typically classify terrestrial ecosystems as either a landcover type [335] or a mixture of PFTs [336]. A savanna is an example of an ecosystem that is a mixture of grass and tree PFTs. Models based on a landcover classification have parameterizations that are intended to represent the weighted average for all of the vegetation species found in the biome. Plant functional types represent groups of vegetation species that are similar for at least some physiological and ecological traits. While it is possible for biome schemes to have very detailed classes, those used in global land surface models are simple approaches that provide a limited ability to represent BVOC emission diversity. A scheme with just five vegetation types (e.g., broadleaf forests, needleleaf forests, shrublands, grasslands, and croplands) was able to account for a significant part of BVOC emission diversity [337]. A small to moderate number (5 to 25) of global PFTs provide a reasonable approach for estimating global isoprene emissions at coarse resolution but cannot represent the considerable within-biome emission diversity which results in large errors in local to regional isoprene emission estimates [5].

The MEGAN2.1 [5] approach for simulating BVOC emission diversity is based on the Community Land Model version 4 PFT scheme [338]. The CLM4 approach is typical of the PFT schemes used for the land surface component of global earth system models and includes 6 temperate, 5 boreal/arctic, 3 tropical, and 1 crop PFTs. Table 5 outlines a framework to improve BVOC emission model estimates by expanding the 15 CLM PFTs to 39 PFTs that can better represent the biological BVOC diversity in earth system models. This approach includes a representative "type" species for each of the PFTs listed in Table 5. The first step towards implementing this approach is to conduct an extensive and systematic quantification of the BVOC emission rates of each of these species. This can be accomplished with enclosure measurements [129] or above-canopy flux measurements above monospecific stands [194]. Additional PFTs can be added when it can be demonstrated that their emission characteristics are substantially different from those on this list.

Table 5: Plant functional type (PFT) scheme for representing BVOC emission biological diversity The 15 CLM4 PFTs are subdivided into BVOC emission types (BVOC-PFT). The measured emission characteristics of the indicated representative species can be used to assign BVOC emission factors

CLM4 PFT description	BVOC-PFT	Representative species
Needleleaf evergreen temperate tree	NETT-IM	Picea engelmannii (Engelmann spruce)
Needleleaf evergreen temperate tree	NETT-MT	Abies grandis (grand fir)
Needleleaf evergreen temperate tree	NETT-MBO	Pinus ponderosa (ponderosa pine)
Needleleaf evergreen temperate tree	NETT-Low	Thuja plicata (western redcedar)
Needleleaf evergreen boreal tree	NEBT-MBO	Pinus contorta (lodgepole pine)
Needleleaf evergreen boreal tree	NEBT-Iso	Picea mariana (black spruce)
Needleleaf evergreen boreal tree	NEBT-MT	Pinus sylvestris (Scots pine)
Needleleaf deciduous boreal tree	NDBT-MT	Larix sibirica (Siberian larch)
Broadleaf evergreen temperate tree	BETE-Iso	Quercus virginiana (southern live oak)
Broadleaf evergreen temperate tree	BETE-IM	Eucalyptus globulus (blue gum)

Broadleaf evergreen temperate tree	BETE-MT	Quercus ilex (holm oak)
Broadleaf evergreen temperate tree	BETE-Low	Lithocarpus densiflorus (tanoak)
Broadleaf evergreen tropical tree	BETR-Iso	Mangifera indica (mango)
Broadleaf evergreen tropical tree	BETR-MT	Tropical evergreen MT emitters
Broadleaf evergreen tropical tree	BETR-Low	Panda oleosa
Broadleaf deciduous tropical tree	BDTR-Iso	Hymenaea courbaril (jatobá)
Broadleaf deciduous tropical tree	BDTR-MT	Apeiba tibourbou
Broadleaf deciduous tropical tree	BDTR-Low	Combretum molle (velvet bushwillow)
Broadleaf deciduous temperate tree	BDTE-Iso	Quercus rubra (red oak)
Broadleaf deciduous temperate tree	BDTE-IM	Liquidambar styraciflua (sweetgum)
Broadleaf deciduous temperate tree	BDTE-MT	Acer saccharum (sugar maple)
Broadleaf deciduous temperate tree	BDTE-Low	Sassafras albidum (sassafras)
Broadleaf deciduous boreal tree	BDBT-Iso	Populus tremuloides (aspen)
Broadleaf deciduous boreal tree	BDBT- Low	Betula pendula (silver birch)
Broadleaf evergreen temperate shrub	BETS-MT	Larrea tridentata (creosote bush)
Broadleaf evergreen temperate shrub	BETS-Iso	Karwinskia humboldtiana (coyotillo)
Broadleaf evergreen temperate shrub	BETS-Low	Atriplex canescens (four-wing saltbush)
Broadleaf deciduous temperate shrub	BDTS-MT	Ambrosia dumosa (white bursage)
Broadleaf deciduous temperate shrub	BDTS-Iso	Psorothamnus fremontii (Fremont's dalea)

Broadleaf deciduous temperate shrub	BDTS-Low	Baccharis texana (prairie false willow)
Broadleaf deciduous boreal shrub	BDBS-Iso	Salix arctica (arctic willow)
Broadleaf deciduous boreal shrub	BDBS-Low	Alnus crispa (mountain alder)
C3 grass	C3G-Iso	Carex appendiculata (sedge)
C3 grass	C3G-Low	Koeleria cristata (June Grass)
C4 Grass	C4G-low	Bouteloua curtipendula (sideoats grama)
Arctic C3 grass	AC3G-Iso	Eriophorum angustifolium (cottongrass)
Arctic C3 grass	AC3G-Low	Festuca rubra (red fescue)
Crop	CRP-MT	Helianthus annuus (sunflower)
Crop	CRP-Iso	Mucuna pruriens (velvet bean)
Crop	CRP-Low	Triticum aestivum (wheat)

The three needleleaf tree PFTs included in CLM4 are temperate evergreen, boreal evergreen, and boreal deciduous. Figure 3 shows that needleleaf trees cover about 15% of the global vegetation covered land area but are estimated to contribute less than 5% of the total BVOC. Figure 3 also shows that nearly 25% of BVOC studies have targeted needleleaf trees indicating that they are relatively well studied. The studies summarized in Table 4 show that all three PFTs include a monoterpene emitting type. Both temperate and boreal evergreen species also include high isoprene [251] and high MBO [68] emitters, and there is some indication that there should be a low emission category for at least the temperate evergreen trees [284]. It should be noted that the available data for characterizing emissions is limited, and the results of different studies are often conflicting. For example, a literature review [284] indicated that the Pseudotsuga menziesii (Douglas fir) monoterpene emission factor is a factor of 8 higher than that of Tsuga heterophylla (western hemlock). In contrast, another study [147] found that the western hemlock monoterpene emission factor is more than twice as high as the value for Douglas fir.

The CLM4 PFTs for broadleaf trees include tropical evergreen, tropical deciduous, temperate evergreen, temperate deciduous, and boreal deciduous trees. These broadleaf trees cover about a third of the vegetation-covered earth surface and are estimated to account for almost 80% of the global total BVOC emission (Figure3). About half of the BVOC emission diversity studies in Tables 1 to 3 have focused on broadleaf trees resulting in a relatively good characterization of temperate and boreal species, but tropical broadleaf tree emissions have not received enough attention (Figure 2). Each of the five CLM4 broadleaf tree PFTs (Table 5) include a high isoprene emitting type. Some also include a high MT emission type [241, 339], a high isoprene and high monoterpene emission type [181], and a low emission type [284].

The CLM4 scheme includes just three shrub PFTs: broadleaf deciduous temperate, broadleaf evergreen temperate and broadleaf deciduous boreal. The two temperate shrub PFTs include high monoterpene, high isoprene, and low emitter categories [65, 138, 340, 341]. The boreal shrub PFT includes both high isoprene and low emitters [256].

The three CLM4 grass PFTs are C3 grass, C4 grass, and arctic C3 grass. All three PFTs are dominated by a low terpenoid emitting category, but the temperate and arctic C3 PFTs also include some isoprene emitting species [22, 49, 100, 233, 342]. The crop PFT is dominated by low terpenoid emitters, but there are some examples of high isoprene and high monoterpene emitters [275, 278, 280, 343].

CONCLUSIONS

This review summarizes the current understanding of BVOC chemical and biological diversity. There are hundreds of BVOCs emitted into the atmosphere, but a relatively few compounds (e.g., isoprene, methanol, α-pinene, acetone, and ethene) dominate the total flux. All BVOCs can influence atmospheric composition, if they are emitted at sufficient rates, but some BVOCs have a relatively high impact due to their reaction rates, products, ozone production potentials, organic aerosol yields, and other properties. As a result, there is a strong need to quantify the chemical diversity of BVOC emissions. On the other hand, a detailed numerical description of BVOC chemical speciation increases computational requirements and the personnel needed to process input variables. In addition, the large uncertainties associated with BVOC emission estimates do not justify an overly detailed parameterization of these compounds. An approach for accurately representing BVOC chemical diversity in emission models requires a balance between providing the appropriate level of details while also minimizing the complexity.

Global land surface models simulate regional variations in ecosystem-atmosphere carbon exchange by assigning values of the photosynthetic parameter V_{cmax} to each PFT. This parameter describes the maximum rate of carboxylation by the photosynthetic enzyme Rubisco. The values of V_{cmax} assigned to the 15 PFTs used by CLM4 vary from 52 μmol m^{-2} s^{-1} for grasses to 72 μmol m^{-2} s^{-1} for broadleaf evergreen trees and shrubs [18]. In contrast, the isoprene emission factor, which describes the isoprene emission rate at a set of standard conditions, ranges from 1 μg m^{-2} h^{-1} for boreal deciduous needleleaf trees to 11000 μmol m^{-2} s^{-1} for broadleaf deciduous boreal trees [5]. This comparison illustrates that there is a much greater range in the ability of plants to emit isoprene than there is for photosynthesis. Assigning BVOC emission factors to 15 PFTs is a good initial step towards characterizing BVOC biological diversity, but it is insufficient. A scheme with about 39 PFTs is proposed to improve regional to global BVOC emission estimates.

Reducing uncertainties in BVOC emission estimates will require additional observations. Measurements are especially needed for specific vegetation types (e.g., tropical broadleaf forest and crops) and some nonterpenoid compounds (e.g., ethene, propene, ethanol, ocimene, and hexenal). Leaf-level enclosure measurements are needed to improve representations of the processes controlling emission variations. Tower- and aircraft-based above-canopy flux measurements are also needed to quantify BVOC diversity on landscape to regional scales.

REFERENCES

1.   T. Pierce, C. Geron, L. Bender, R. Dennis, G. Tonnesen, and A. Guenther, "Influence of increased isoprene emissions on regional

ozone modeling," Journal of Geophysical Research D, vol. 103, no. 19, pp. 25611–25629, 1998.

2.   A. G. Carlton, R. W. Pinder, P. V. Bhave, and G. A. Pouliot, "To what extent can biogenic SOA be controlled?" Environmental Science and Technology, vol. 44, no. 9, pp. 3376–3380, 2010. · ·

3.   D. V. Spracklen, J. L. Jimenez, K. S. Carslaw et al., "Aerosol mass spectrometer constraint on the global secondary organic aerosol budget," Atmospheric Chemistry and Physics, vol. 11, no. 23, pp. 12109–12136, 2011. · ·

4.   R. Baghi, D. Helmig, A. Guenther, T. Duhl, and R. Daly, "Contribution of flowering trees to urban atmospheric biogenic volatile organic compound emissions," Biogeosciences, vol. 9, pp. 3777–3785, 2012.

5.   A. B. Guenther, X. Jiang, C. L. Heald et al., "The model of emissions of gases and aerosols from nature version 2. 1 (MEGAN2. 1): an extended and updated framework for modeling biogenic emissions," Geoscientific Model Development, vol. 5, no. 6, pp. 1471–1492, 2012.

6.   J. P. Greenberg, D. Asensio, A. Turnipseed, A. B. Guenther, T. Karl, and D. Gochis, "Contribution of leaf and needle litter to whole ecosystem BVOC fluxes," Atmospheric Environment, vol. 59, pp. 302–311, 2012.

7.   T. E. Graedel, "Terpenoids in the atmosphere," Reviews of Geophysics & Space Physics, vol. 17, no. 5, pp. 937–948, 1979.

8.   A. H. Goldstein and I. E. Galbally, "Known and unexplored organic constituents in the earth›s atmosphere," Environmental Science and Technology, vol. 41, no. 5, pp. 1514–1521, 2007. · ·

9.   R. A. Rasmussen, "What do the hydrocarbons from trees contribute to air pollution?" Journal of the Air Pollution Control Association, vol. 22, no. 7, pp. 537–543, 1972.

10. R. Singh, A. P. Singh, M. P. Singh, A. Kumar, and C. K. Varshney, "Emission of isoprene from common Indian plant species and its implications for regional air quality," Environmental Monitoring and Assessment, vol. 144, no. 1–3, pp. 43–51, 2008. · ·

11. P. Ciccioli, E. Brancaleoni, M. Frattoni et al., "Emission of reactive terpene compounds from orange orchards and their removal by within-canopy processes," Journal of Geophysical Research D, vol. 104, no. 7, pp. 8077–8094, 1999.

12. N. Dudareva, F. Negre, D. A. Nagegowda, and I. Orlova, "Plant volatiles: recent advances and future perspectives," Critical Reviews in Plant Sciences, vol. 25, no. 5, pp. 417–440, 2006. · ·

13. F. W. Went, "Blue hazes in the atmosphere," Nature, vol. 187, no. 4738, pp. 641–643, 1960. · ·

14. P. Zimmerman, "Testing of hydrocarbon emissions from vegetastion, leaf litter and aquatic surfaces and devlopment of a method for compiling biogenic emission inventories," Tech. Rep. EPA-450-4-70-004, U.S. Environmental Protection Agency, Research Triangle Park, calif, USA, 1979.

15. B. Lamb, A. Guenther, D. Gay, and H. Westberg, "A national inventory of biogenic hydrocarbon emissions," Atmospheric Environment, vol. 21, no. 8, pp. 1695–1705, 1987.

16. J.-F. Muller, "Geographical distribution and seasonal variation of surface emissions and deposition velocities of atmospheric trace gases," Journal of Geophysical Research, vol. 97, no. 4, pp. 3787–3804, 1992.

17. A. Guenther, C. N. Hewitt, D. Erickson, et al., "A global model of natural volatile organic compound emissions," Journal of geophysical research, vol. 100, no. 5, pp. 8873–8892, 1995.

18. D. M. Lawrence, K. W. Oleson, M. G. Flanner, et al., "Parameterization improvements and functional and structural advances in version 4 of the community land model," Journal of Advances in Modeling Earth Systems, vol. 3, 27 pages, 2011.

19. E. C. Apel, D. D. Riemer, A. Hills et al., "Measurement and interpretation of isoprene fluxes isoprene, methacrolein, and methyl vinyl ketone mixing ratios at the PROPHET site during the 1998 intensive,"Journal of Geophysical Research D, vol. 107, no. 3, pp. 1–15, 2002.

20. R. R. Arnts, W. B. Petersen, R. L. Seila, and B. W. Gay Jr., "Estimates of α-pinene emissions from a loblolly pine forest using

an atmospheric diffusion model," Atmospheric Environment, vol. 16, no. 9, pp. 2127–2137, 1982.

21. G. P. Ayers and R. W. Gillett, "Isoprene emissions from vegetation and hydrocarbon emissions from bushfires in tropical Australia," Journal of Atmospheric Chemistry, vol. 7, no. 2, pp. 177–190, 1988.

22. J. Bai, B. Baker, B. Liang, J. Greenberg, and A. Guenther, "Isoprene and monoterpene emissions from an Inner Mongolia grassland," Atmospheric Environment, vol. 40, no. 30, pp. 5753–5758, 2006. · ·

23. B. Baker, A. Guenther, J. Greenberg, A. Goldstein, and R. Fall, "Canopy fluxes of 2-methyl-3-buten-2-ol over a ponderosa pine forest by relaxed eddy accumulation: field data and model comparison," Journal of Geophysical Research D, vol. 104, no. 21, pp. 26107–26114, 1999.

24. B. Baker, A. Guenther, J. Greenberg, and R. Fall, "Canopy level fluxes of 2-methyl-3-buten-2-ol, acetone, and methanol by a portable relaxed eddy accumulation system," Environmental Science and Technology, vol. 35, no. 9, pp. 1701–1708, 2001. · ·

25. I. Bamberger, L. Hörtnagl, T. M. Ruuskanen et al., "Deposition fluxes of terpenes over grassland,"Journal of Geophysical Research D, vol. 116, no. 14, Article ID D14305, 2011. · ·

26. J. N. Barney, J. P. Sparks, J. Greenberg, T. H. Whitlow, and A. Guenther, "Biogenic volatile organic compounds from an invasive species: impacts on plant-plant interactions," Plant Ecology, vol. 203, no. 2, pp. 195–205, 2009. · ·

27. B. Bonsang, G. K. Moortgat, and C. A. Pio, "Overview of the FIELDVOC›94 experiment in a eucalyptus forest of Portugal," Chemosphere, vol. 3, no. 3, pp. 211–226, 2001. · ·

28. J. W. Bottenheim and M. F. Shepherd, "C_2-C_6 hydrocarbon measurements at four rural locations across Canada,"Atmospheric Environment, vol. 29, no. 6, pp. 647–664, 1995. · ·

29. A. Bracho-Nunez, S. Welter, M. Staudt, and J. Kesselmeier, "Plant-specific volatile organic compound emission rates from young and mature leaves of Mediterranean vegetation," Journal

of Geophysical Research D, vol. 116, no. 16, Article ID D16304, 2011. · ·

30. P. T. Buckley, "Isoprene emissions from a Florida scrub oak species grown in ambient and elevated carbon dioxide," Atmospheric Environment, vol. 35, no. 3, pp. 631–634, 2001. · ·

31. K. E. Burr, S. J. Wallner, and R. W. Tinus, "Ethylene and ethane evolution during cold acclimation and deacclimation of ponderosa pine," Canadian Journal of Forestry Research, vol. 21, pp. 601–605, 1991.

32. C. Calfapietra, G. Scarascia Mugnozza, D. F. Karnosky, F. Loreto, and T. D. Sharkey, "Isoprene emission rates under elevated CO_2 and O_3 in two field-grown aspen clones differing in their sensitivity to O_3,"New Phytologist, vol. 179, no. 1, pp. 55–61, 2008. · ·

33. X.-L. Cao, C. Boissard, A. J. Juan, C. N. Hewitt, and M. Gallagher, "Biogenic emissions of volatile organic compounds from gorse (Ulex europaeus): diurnal emission fluxes at Kelling Heath, England,"Journal of Geophysical Research D, vol. 102, no. 15, pp. 18903–18915, 1997.

34. J. Chang, Y. Ren, Y. Shi, et al., "An inventory of biogenic volatile organic compounds for a subtropical urban-rural complex," Atmospheric Environment, vol. 56, pp. 115–123, 2012.

35. P. Ciccioli, C. Fabozzi, E. Brancaleoni et al., "Biogenic emission from the Mediterranean Pseudosteppe ecosystem present in Castelporziano," Atmospheric Environment, vol. 31, no. 1, pp. 167–175, 1997. · ·

36. S. B. Corchnoy, J. Arey, and R. Atkinson, "Hydrocarbon emissions from twelve urban shade trees of the Los Angeles, California, Air Basin," Atmospheric Environment, vol. 26, no. 3, pp. 339–348, 1992. · ·

37. B. Davison, R. Taipale, B. Langford et al., "Concentrations and fluxes of biogenic volatile organic compounds above a Mediterranean Macchia ecosystem in western Italy," Biogeosciences, vol. 6, no. 8, pp. 1655–1670, 2009.

38. W. A. Dement, B. J. Tyson, and H. A. Mooney, "Mechanism of monoterpene volatilization in Salvia mellifera," Phytochemistry, vol. 14, no. 12, pp. 2555–2557, 1975.

39. J. P. Digangi, E. S. Boyle, T. Karl et al., "First direct measurements of formaldehyde flux via eddy covariance: implications for missing in-canopy formaldehyde sources," Atmospheric Chemistry and Physics, vol. 11, no. 20, pp. 10565–10578, 2011. · ·

40. T. Dindorf, U. Kuhn, L. Ganzeveld et al., "Significant light and temperature dependent monoterpene emissions from European beech (Fagus sylvatica L.) and their potential impact on the European volatile organic compound budget," Journal of Geophysical Research D, vol. 111, no. 16, Article ID D16305, 2006. · ·

41. P. Dominguez-Taylor, L. G. Ruiz-Suarez, I. Rosas-Perez, J. M. Hernández-Solis, and R. Steinbrecher, "Monoterpene and isoprene emissions from typical tree species in forests around Mexico City,"Atmospheric Environment, vol. 41, no. 13, pp. 2780–2790, 2007. · ·

42. P. V. Doskey and W. Gao, "Vertical mixing and chemistry of isoprene in the atmospheric boundary layer: aircraft-based measurements and numerical modeling," Journal of Geophysical Research D, vol. 104, no. 17, pp. 21263–21274, 1999.

43. R. C. Evans, D. T. Tingey, M. L. Gumpertz, and W. F. Burns, "Estimates of isoprene and monoterpene emission rates in plants," Botanical Gazette, vol. 143, no. 3, pp. 304–310, 1982.

44. S. Fares, J. H. Park, D. R. Gentner et al., "Seasonal cycles of biogenic volatile organic compound fluxes and concentrations in a California citrus orchard," Atmospheric Chemistry and Physics, vol. 12, no. 20, pp. 9865–9880, 2012.

45. P. Fruekilde, J. Hjorth, N. R. Jensen, D. Kotzias, and B. Larsen, "Ozonolysis at vegetation surfaces: a source of acetone, 4-oxopentanal, 6-methyl-5-hepten-2-one, and geranyl acetone in the troposphere,"Atmospheric Environment, vol. 32, no. 11, pp. 1893–1902, 1998. · ·

46. J. D. Fuentes, D. Wang, G. Den Hartog, H. H. Neumann, T. F. Dann, and K. J. Puckett, "Modelled and field measurements of biogenic hydrocarbon emissions from a Canadian deciduous forest," Atmospheric Environment, vol. 29, no. 21, pp. 3003–3017, 1995.

47. J. D. Fuentes, D. Wang, H. H. Neumann, T. J. Gillespie, G. Den Hartog, and T. F. Dann, "Ambient biogenic hydrocarbons and isoprene emissions from a mixed deciduous forest," Journal of Atmospheric Chemistry, vol. 25, no. 1, pp. 67–95, 1996.

48. J. D. Fuentes and D. Wang, "On the seasonality of isoprene emissions from a mixed temperate forest," Ecological Applications, vol. 9, no. 4, pp. 1118–1131, 1999.

49. Y. Fukui and P. V. Doskey, "Air-surface exchange of nonmethane organic compounds at a grassland site: seasonal variations and stressed emissions," Journal of Geophysical Research D, vol. 103, no. 11, pp. 13153–13168, 1998.

50. F. Geng, X. Tie, A. Guenther, G. Li, J. Cao, and P. Harley, "Effect of isoprene emissions from major forests on ozone formation in the city of Shanghai, China," Atmospheric Chemistry and Physics, vol. 11, no. 20, pp. 10449–10459, 2011.

51. C. D. Geron, D. Nie, R. R. Arnts et al., "Biogenic isoprene emission: model evaluation in a southeastern United States bottomland deciduous forest," Journal of Geophysical Research D, vol. 102, no. 15, pp. 18889–18901, 1997.

52. C. Geron, A. Guenther, T. Sharkey, and R. R. Arnts, "Temporal variability in basal isoprene emission factor," Tree Physiology, vol. 20, no. 12, pp. 799–805, 2000.

53. C. Geron, R. Rasmussen, R. R. Arnts, and A. Guenther, "A review and synthesis of monoterpene speciation from forests in the United States," Atmospheric Environment, vol. 34, no. 11, pp. 1761–1781, 2000.

54. C. Geron, P. Harley, and A. Guenther, "Isoprene emission capacity for US tree species," Atmospheric Environment, vol. 35, no. 19, pp. 3341–3352, 2001.

55. P. D. Goldan, W. C. Kuster, F. C. Fehsenfield, and S. A. Montzka, "The observation of a C_5 alcohol emission in a north American pine forest," Geophysical Research Letters, vol. 20, no. 11, pp. 1039–1042, 1993.

56. A. H. Goldstein, S. M. Fan, M. L. Goulden, J. W. Munger, and S. C. Wofsy, "Emissions of ethene, propene, and 1-butene by a midlatitude forest," Journal of Geophysical Research D, vol. 101, no. 4 D, pp. 9149–9157, 1996.

57. A. H. Goldstein, M. L. Goulden, J. W. Munger, S. C. Wofsy, and C. D. Geron, "Seasonal course of isoprene emissions from a midlatitude deciduous forest," Journal of Geophysical Research D, vol. 103, no. 23, pp. 31045–31056, 1998.

58. J. P. Greenberg, A. Guenther, P. Zimmerman et al., "Tethered balloon measurements of biogenic VOCs in the atmospheric boundary layer," Atmospheric Environment, vol. 33, no. 6, pp. 855–867, 1999. · ·

59. A. B. Guenther, R. K. Monson, and R. Fall, "Isoprene and monoterpene emission rate variability: observations with eucalyptus and emission rate algorithm development," Journal of Geophysical Research, vol. 96, no. 6, pp. 10799–10808, 1991.

60. A. B. Guenther, P. R. Zimmerman, P. C. Harley, R. K. Monson, and R. Fall, "Isoprene and monoterpene emission rate variability: model evaluations and sensitivity analyses," Journal of Geophysical Research, vol. 98, no. 7, pp. 12–617, 1993.

61. A. B. Guenther and A. J. Hills, "Eddy covariance measurement of isoprene fluxes," Journal of Geophysical Research D, vol. 103, no. 11, pp. 13145–13152, 1998.

62. A. Guenther, J. Greenberg, P. Harley et al., "Leaf, branch, stand and landscape scale measurements of volatile organic compound fluxes from U.S. woodlands," Tree Physiology, vol. 16, no. 1-2, pp. 17–24, 1996.

63. A. Guenther, P. Zimmerman, L. Klinger, et al., "Estimates of regional natural volatile organic compound fluxes from enclosure and ambient measurements," Journal of Geophysical Research, vol. 101, no. 1, pp. 1345–1359, 1966.

64. A. Guenther, W. Baugh, K. Davis et al., "Isoprene fluxes measured by enclosure, relaxed eddy accumulation, surface layer gradient, mixed layer gradient, and mixed layer mass balance techniques,"Journal of Geophysical Research D, vol. 101, no. 13, pp. 18555–18567, 1996.

65. A. Guenther, S. Archer, J. Greenberg et al., "Biogenic hydrocarbon emissions and landcover/climate change in a subtropical savanna," Physics and Chemistry of the Earth B, vol. 24, no. 6, pp. 659–667, 1999. · ·

66. P. Harley, A. Guenther, and P. Zimmerman, "Effects of light, temperature and canopy position on net photosynthesis and

isoprene emission from sweetgum (Liquidambar styraciflua) leaves," Tree Physiology, vol. 16, no. 1-2, pp. 25–32, 1996.

67. P. Harley, A. Guenther, and P. Zimmerman, "Environmental controls over isoprene emission in deciduous oak canopies," Tree Physiology, vol. 17, no. 11, pp. 705–714, 1997.

68. P. Harley, V. Fridd-Stroud, J. Greenberg, A. Guenther, and P. Vasconcellos, "Emission of 2-methyl-3-buten-2-ol by pines: a potentially large natural source of reactive carbon to the atmosphere," Journal of Geophysical Research D, vol. 103, no. 19, pp. 25479–25486, 1998.

69. P. Harley, L. Otter, A. Guenther, and J. Greenberg, "Micrometeorological and leaf-level measurements of isoprene emissions from a southern African savanna," Journal of Geophysical Research, vol. 108, no. 13, 2003. · ·

70. P. Harley, J. Greenberg, Ü. Niinemets, and A. Guenther, "Environmental controls over methanol emission from leaves," Biogeosciences, vol. 4, no. 6, pp. 1083–1099, 2007.

71. D. Harrison, M. C. Hunter, A. C. Lewis, P. W. Seakins, T. V. Nunes, and C. A. Pio, "Isoprene and monoterpene emission from the coniferous species Abies Borisii-regis: implications for regional air chemistry in Greece," Atmospheric Environment, vol. 35, no. 27, pp. 4687–4698, 2001. · ·

72. C. He, F. Murray, and T. Lyons, "Seasonal variations in monoterpene emissions from Eucalyptus species," Chemosphere, vol. 2, no. 1, pp. 65–76, 2000. · ·

73. A. C. Heiden, T. Hoffmann, J. Kahl et al., "Emission of volatile organic compounds from ozone-exposed plants," Ecological Applications, vol. 9, no. 4, pp. 1160–1167, 1999.

74. D. Helmig, L. F. Klinger, A. Guenther, L. Vierling, C. Geron, and P. Zimmerman, "Biogenic volatile organic compound emissions (BVOCs). I. Identifications from three continental sites in the U.S.,"Chemosphere, vol. 38, no. 9, pp. 2163–2187, 1999. · ·

75. D. Helmig, J. Ortega, A. Guenther, J. D. Herrick, and C. Geron, "Sesquiterpene emissions from loblolly pine and their potential contribution to biogenic aerosol formation in the Southeastern US,"Atmospheric Environment, vol. 40, no. 22, pp. 4150–4157, 2006. · ·

76. D. Helmig, J. Ortega, T. Duhl et al., "Sesquiterpene emissions from pine trees: identifications, emission rates and flux estimates for the contiguous United States," Environmental Science and Technology, vol. 41, no. 5, pp. 1545–1553, 2007. · ·

77. M. W. Holdren, H. H. Westberg, and P. R. Zimmerman, "Analysis of monoterpene hydrocarbons in rural atmosphere," Journal of Geophysical Research, vol. 84, no. 8, pp. 5083–5088, 1979.

78. R. Holzinger, A. Lee, M. McKay, and A. H. Goldstein, "Seasonal variability of monoterpene emission factors for a Ponderosa pine plantation in California," Atmospheric Chemistry and Physics, vol. 6, no. 5, pp. 1267–1274, 2006.

79. C. Holzke, T. Dindorf, J. Kesselmeier, U. Kuhn, and R. Koppmann, "Terpene emissions from European beech (Fagus sylvatica L.): pattern and emission behaviour over two vegetation periods," Journal of Atmospheric Chemistry, vol. 55, no. 1, pp. 81–102, 2006. · ·

80. A.-K. Huang, N. Li, A. Guenther et al., "Investigation on emission properties of biogenic VOCs of landscape plants in Shenzhen," Huanjing Kexue/Environmental Science, vol. 32, no. 12, pp. 3555–3559, 2011.

81. J. G. Isebrands, A. B. Guenther, P. Harley et al., "Volatile organic compound emission rates from mixed deciduous and coniferous forests in Northern Wisconsin, USA," Atmospheric Environment, vol. 33, no. 16, pp. 2527–2536, 1999. · ·

82. V. A. Isidorov, I. G. Zenkevich, and B. V. Ioffe, "Volatile organic compounds in the atmosphere of forests," Atmospheric Environment, vol. 19, no. 1, pp. 1–8, 1985.

83. K. Jardine, T. Karl, M. Lerdau, P. Harley, A. Guenther, and J. E. Mak, "Carbon isotope analysis of acetaldehyde emitted from leaves following mechanical stress and anoxia," Plant Biology, vol. 11, no. 4, pp. 591–597, 2009. · ·

84. K. Jardine, P. Harley, T. Karl, A. Guenther, M. Lerdau, and J. E. Mak, "Plant physiological and environmental controls over the exchange of acetaldehyde between forest canopies and the atmosphere," Biogeosciences, vol. 5, no. 6, pp. 1559–1572, 2008.

85. K. J. Jardine, E. D. Sommer, S. R. Saleska, T. E. Huxman, P. C. Harley, and L. Abrell, "Gas phase measurements of pyruvic

acid and its volatile metabolites," Environmental Science and Technology, vol. 44, no. 7, pp. 2454–2460, 2010. ·

86. T. G. Karl, C. Spirig, J. Rinne et al., "Virtual disjunct eddy covariance measurements of organic compound fluxes from a subalpine forest using proton transfer reaction mass spectrometry,"Atmospheric Chemistry and Physics, vol. 2, no. 4, pp. 279–291, 2002.

87. T. Karl, A. Guenther, C. Spirig, A. Hansel, and R. Fall, "Seasonal variation of biogenic VOC emissions above a mixed hardwood forest in northern Michigan," Geophysical Research Letters, vol. 30, no. 23, pp. 4–19, 2003.

88. T. Karl, A. Guenther, A. Turnipseed, E. G. Patton, and K. Jardine, "Chemical sensing of plant stress at the ecosystem scale," Biogeosciences, vol. 5, no. 5, pp. 1287–1294, 2008.

89. J. F. Karlik and A. M. Winer, "Measured isoprene emission rates of plants in California landscapes: comparison to estimates from taxonomic relationships," Atmospheric Environment, vol. 35, no. 6, pp. 1123–1131, 2001.

90. J. Kesselmeier, L. Schäfer, P. Ciccioli et al., "Emission of monoterpenes and isoprene from a Mediterranean oak species Quercus ilex L. measured within the BEMA (Biogenic Emissions in the Mediterranean Area) project," Atmospheric Environment, vol. 30, no. 10-11, pp. 1841–1850, 1996. · ·

91. J. Kesselmeier, K. Bode, U. Hofmann et al., "Emission of short chained organic acids, aldehydes and monoterpenes from Quercus ilex L. and Pinus pinea L. in relation to physiological activities, carbon budget and emission algorithms," Atmospheric Environment, vol. 31, no. 1, pp. 119–133, 1997.

92. J. Kesselmeier, K. Bode, L. Schafer et al., "Simultaneous field measurements of terpene and isoprene emissions from two dominant Mediterranean oak species in relation to a North American species,"Atmospheric Environment, vol. 32, no. 11, pp. 1947–1953, 1998. · ·

93. J. Kesselmeier, K. Bode, C. Gerlach, and E.-M. Jork, "Exchange of atmospheric formic and acetic acids with trees and crop plants under controlled chamber and purified air conditions," Atmospheric Environment, vol. 32, no. 10, pp. 1765–1775, 1998. · ·

94. J.-C. Kim, "Factors controlling natural VOC emissions in a southeastern US pine forest," Atmospheric Environment, vol. 35, no. 19, pp. 3279–3292, 2001. · ·

95. J.-C. Kim, K.-J. Kim, D.-S. Kim, and J.-S. Han, "Seasonal variations of monoterpene emissions from coniferous trees of different ages in Korea," Chemosphere, vol. 59, no. 11, pp. 1685–1696, 2005.

96. S. Kim, T. Karl, D. Helmig, R. Daly, R. Rasmussen, and A. Guenther, "Measurement of atmospheric sesquiterpenes by proton transfer reaction-mass spectrometry (PTR-MS)," Atmospheric Measurement Techniques, vol. 2, no. 1, pp. 99–112, 2009.

97. S. Kim, T. Karl, A. Guenther et al., "Emissions and ambient distributions of Biogenic Volatile Organic Compounds (BVOC) in a ponderosa pine ecosystem: interpretation of PTR-MS mass spectra,"Atmospheric Chemistry and Physics, vol. 10, no. 4, pp. 1759–1771, 2010.

98. S. Kim, A. Guenther, T. Karl, and J. Greenberg, "Contributions of primary and secondary biogenic VOC tototal OH reactivity during the CABINEX (Community Atmosphere-Biosphere INteractions Experiments)-09 field campaign," Atmospheric Chemistry and Physics, vol. 11, no. 16, pp. 8613–8623, 2011.

99. S. Y. Kim, X. Y. Jiang, M. Lee, et al., "Impact of biogenic volatile organic compounds on ozone production at the Taehwa Research Forest near Seoul, South Korea," Atmospheric Environment, vol. 70, pp. 447–453, 2013.

100. W. Kirstine, I. Galbally, Y. Ye, and M. Hooper, "Emissions of volatile organic compounds (primarily oxygenated species) from pasture," Journal of Geophysical Research D, vol. 103, no. 3339, pp. 10605–10619, 1998.

101. L. F. Klinger, Q. J. Li, A. B. Guenther, J. P. Greenberg, B. Baker, and J. H. Bai, "Assessment of volatile organic compound emissions from ecosystems of China," Journal of Geophysical Research, vol. 107, no. 21, 2002.

102. G. Konig, M. Brunda, H. Puxbaum, C. N. Hewitt, S. C. Duckham, and J. Rudolph, "Relative contribution of oxygenated hydrocarbons to the total biogenic VOC emissions of selected mid-European agricultural and natural plant species," Atmospheric Environment, vol. 29, no. 8, pp. 861–874, 1995. · ·

103. J. Kreuzwieser, J.-P. Schnitzler, and R. Steinbrecher, "Biosynthesis of organic compounds emitted by plants," Plant Biology, vol. 1, no. 2, pp. 149–159, 1999.

104. J. Kreuzwieser, H. Rennenberg, and R. Steinbrecher, "Impact of short-term and long-term elevated CO_2 on emission of carbonyls from adult Quercus petraea and Carpinus betulus trees," Environmental Pollution, vol. 142, no. 2, pp. 246–253, 2006. ·
·

105. B. Lamb, H. Westberg, and G. Allwine, "Biogenic hydrocarbon emissions from deciduous and coniferous trees in the United States," Journal of Geophysical Research, vol. 90, no. 1, pp. 2380–2390, 1985.

106. B. Lamb, H. Westberg, and G. Allwine, "Isoprene emission fluxes determined by an atmospheric tracer technique," Atmospheric Environment, vol. 20, no. 1, pp. 1–8, 1986.

107. M. Lerdau, S. B. Dilts, H. Westberg, B. K. Lamb, and E. J. Allwine, "Monoterpene emission from ponderosa pine," Journal of Geophysical Research, vol. 99, no. 8, pp. 16–615, 1994.

108. M. Lerdau, P. Matson, R. Fall, and R. Monson, "Ecological controls over monoterpene emissions from douglas-fir (Pseudotsuga menziesii)," Ecology, vol. 76, no. 8, pp. 2640–2647, 1995.

109. D. W. Li, Y. Shi, X. Y. He, W. Chen, and X. Chen, "Volatile organic compound emissions from urban trees in Shenyang, China," Botanical Studies, vol. 49, no. 1, pp. 67–72, 2008.

110. Y.-J. Lim, A. Armendariz, Y.-S. Son, and J.-C. Kim, "Seasonal variations of isoprene emissions from five oak tree species in East Asia," Atmospheric Environment, vol. 45, no. 13, pp. 2202–2210, 2011. · ·

111. M. E. Litvak, F. Loreto, P. C. Harley, T. D. Sharkey, and R. K. Monson, "The response of isoprene emission rate and photosynthetic rate to photon flux and nitrogen supply in aspen and white oak trees," Plant, Cell and Environment, vol. 19, no. 5, pp. 549–559, 1996.

112. J. Llusia, J. Penuelas, R. Seco, and I. Filella, "Seasonal changes in the daily emission rates of terpenes byQuercus ilex and the atmospheric concentrations of terpenes in the natural park of Montseny, NE Spain," Journal of Atmospheric Chemistry, vol. 69, no. 3, pp. 215–230, 2012.

113. H. W. Loescher, Non-methane hydrocarbon fluxes from Pinus elliottii and Sereonoa repens: comparing enclosure and above-canopy measurements [Doctoral dissertation], University of Florida, 1997.

114. F. Loreto and T. D. Sharkey, "A gas-exchange study of photosynthesis and isoprene emission in Quercus rubra L," Planta, vol. 182, no. 4, pp. 523–531, 1990

115. R. C. MacDonald and R. Fall, "Detection of substantial emissions of methanol from plants to the atmosphere," Atmospheric Environment, vol. 27, no. 11, pp. 1709–1713, 1993.

116. R. C. MacDonald and R. Fall, "Acetone emission from conifer buds," Phytochemistry, vol. 34, no. 4, pp. 991–994, 1993. · ·

117. M. B. Madronich, J. P. Greenberg, C. A. Wessman, and A. B. Guenther, "Monoterpene emissions from an understory species, Pteridium aquilinum," Atmospheric Environment, vol. 54, pp. 308–312, 2012. · ·

118. R. S. Martin, H. Westberg, E. Allwine, L. Ashman, J. C. Farmer, and B. Lamb, "Measurement of isoprene and its atmospheric oxidation products in a central Pennsylvania deciduous forest," Journal of Atmospheric Chemistry, vol. 13, no. 1, pp. 1–32, 1991.

119. R. S. Martin, I. Villanueva, J. Zhang, and C. J. Popp, "Nonmethane hydrocarbon, monocarboxylic acid, and low molecular weight aldehyde and ketone emissions from vegetation in central New Mexico," Environmental Science and Technology, vol. 33, no. 13, pp. 2186–2192, 1999. · ·

120. S. N. Matsunaga, A. B. Guenther, M. J. Potosnak, and E. C. Apel, "Emission of sunscreen salicylic esters from desert vegetation and their contribution to aerosol formation," Atmospheric Chemistry and Physics, vol. 8, no. 24, pp. 7367–7371, 2008.

121. S. N. Matsunaga, A. B. Guenther, J. P. Greenberg et al., "Leaf level emission measurement of sesquiterpenes and oxygenated sesquiterpenes from desert shrubs and temperate forest trees using a liquid extraction technique," Geochemical Journal, vol. 43, no. 3, pp. 179–189, 2009. · ·

122. S. N. Matsunaga, S. Chatani, S. Nakatsuka et al., "Determination and potential importance of diterpene (kaur-16-ene) emitted from dominant coniferous trees in Japan," Chemosphere, vol. 87, no. 8, pp. 886–893, 2012. · ·

123. S. N. Matsunaga, O. Muller, S. Chatani, M. Nakamura, T. Nakaji, and T. Hiura, "Seasonal variation of isoprene basal emission in mature Quercus crispula trees under experimental warming of roots and branches," Geochemical Journal, vol. 46, no. 2, pp. 163–167, 2012.

124. K. A. McKinney, B. H. Lee, A. Vasta, T. V. Pho, and J. W. Munger, "Emissions of isoprenoids and oxygenated biogenic volatile organic compounds from a New England mixed forest," Atmospheric Chemistry and Physics, vol. 11, no. 10, pp. 4807–4831, 2011. · ·

125. R. K. Monson and R. Fall, "Isoprene Emission from Aspen Leaves: influence of Environment and Relation to Photosynthesis and Photorespiration," Plant Physiology, vol. 90, no. 1, pp. 267–274, 1989.

126. R. K. Monson, P. C. Harley, M. E. Litvak et al., "Environmental and developmental controls over the seasonal pattern of isoprene emission from aspen leaves," Oecologia, vol. 99, no. 3-4, pp. 260–270, 1994.

127. S. Moukhtar, B. Bessagnet, L. Rouil, and V. Simon, "Monoterpene emissions from Beech (Fagus sylvatica) in a French forest and impact on secondary pollutants formation at regional scale," Atmospheric Environment, vol. 39, no. 19, pp. 3535–3547, 2005.

128. Ü. Niinemets, "Mild versus severe stress and BVOCs: thresholds, priming and consequences," Trends in Plant Science, vol. 15, no. 3, pp. 145–153, 2010.

129. Ü. Niinemets, U. Kuhn, P. C. Harley et al., "Estimations of isoprenoid emission capacity from enclosure studies: measurements, data processing, quality and standardized measurement protocols," Biogeosciences, vol. 8, no. 8, pp. 2209–2246, 2011.

130. T. V. Nunes and C. A. Pio, "Emission of volatile organic compounds from Portuguese eucalyptus forests," Chemosphere, vol. 3, no. 3, pp. 239–248, 2001.

131. K. Ohta, "Diurnal and seasonal variation in isoprene emission from live oak," Geochemical Journal, vol. 19, pp. 269–274, 1986.

132. E. Ormeño, C. Fernandez, A. Bousquet-Mélou et al., "Monoterpene and sesquiterpene emissions of three Mediterranean species

through calcareous and siliceous soils in natural conditions," Atmospheric Environment, vol. 41, no. 3, pp. 629–639, 2007.

133. J. Ortega, D. Helmig, A. Guenther, P. Harley, S. Pressley, and C. Vogel, "Flux estimates and OH reaction potential of reactive biogenic volatile organic compounds (BVOCs) from a mixed northern hardwood forest," Atmospheric Environment, vol. 41, no. 26, pp. 5479–5495, 2007.

134. J. Ortega, D. Helmig, R. W. Daly, D. M. Tanner, A. B. Guenther, and J. D. Herrick, "Approaches for quantifying reactive and low-volatility biogenic organic compound emissions by vegetation enclosure techniques. Part B: applications," Chemosphere, vol. 72, no. 3, pp. 365–380, 2008.

135. L. B. Otter, A. Guenther, and J. Greenberg, "Seasonal and spatial variations in biogenic hydrocarbon emissions from southern African savannas and woodlands," Atmospheric Environment, vol. 36, no. 26, pp. 4265–4275, 2002.

136. S. Owen, C. Boissard, R. A. Street, S. C. Duckham, O. Csiky, and C. N. Hewitt, "Screening of 18 Mediterranean plant species for volatile organic compound emissions," Atmospheric Environment, vol. 31, no. 1, pp. 101–117, 1997.

137. S. M. Owen, C. Boissard, B. Hagenlocher, and C. N. Hewitt, "Field studies of isoprene emissions from vegetation in the Northwest Mediterranean region," Journal of Geophysical Research D, vol. 103, no. 19, pp. 25499–25511, 1998.

138. M. R. Papiez, M. J. Potosnak, W. S. Goliff, A. B. Guenther, S. N. Matsunaga, and W. R. Stockwell, "The impacts of reactive terpene emissions from plants on air quality in Las Vegas, Nevada," Atmospheric Environment, vol. 43, no. 27, pp. 4109–4123, 2009.

139. E. Pegoraro, A. Rey, J. Greenberg et al., "Effect of drought on isoprene emission rates from leaves of Quercus virginiana Mill," Atmospheric Environment, vol. 38, no. 36, pp. 6149–6156, 2004.

140. D. Pérez-Rial, J. Peñuelas, P. López-Mahía, and J. Llusià, "Terpenoid emissions from Quercus robur. A case study of Galicia (NW Spain)," Journal of Environmental Monitoring, vol. 11, no. 6, pp. 1268–1275, 2009

141. G. Pétron, P. Harley, J. Greenberg, and A. Guenther, "Seasonal temperature variations influence isoprene emission," Geophysical Research Letters, vol. 28, no. 9, pp. 1707–1710, 2001.

142. P. A. Pier, "Isoprene emission rates from northern red oak using a whole-tree chamber," Atmospheric Environment, vol. 29, no. 12, pp. 1347–1353, 1995.

143. P. A. Pier and C. McDuffie Jr., "Seasonal isoprene emission rates and model comparisons using whole-tree emissions from white oak," Journal of Geophysical Research D, vol. 102, no. 20, pp. 23963–23971, 1997.

144. C. A. Pio and A. A. Valente, "Atmospheric fluxes and concentrations of monoterpenes in resin-tapped pine forests," Atmospheric Environment, vol. 32, no. 4, pp. 683–691, 1998.

145. O. Pokorska, J. Dewulf, C. Amelynck et al., "Isoprene and terpenoid emissions from Abies alba: identification and emission rates under ambient conditions," Atmospheric Environment, vol. 59, pp. 501–508, 2012.

146. O. Pokorska, J. Dewulf, C. Amelynck et al., "Emissions of biogenic volatile organic compounds fromFraxinus excelsior and Quercus robur under ambient conditions in Flanders (Belgium)," International Journal of Environmental Analytical Chemistry, vol. 92, no. 15, pp. 1729–1741, 2012.

147. S. Pressley, B. Lamb, H. Westberg, A. Guenther, J. Chen, and E. Allwine, "Monoterpene emissions from a Pacific Northwest Old-Growth Forest and impact on regional biogenic VOC emission estimates,"Atmospheric Environment, vol. 38, no. 19, pp. 3089–3098, 2004.

148. S. Pressley, B. Lamb, H. Westberg, J. Flaherty, J. Chen, and C. Vogel, "Long-term isoprene flux measurements above a northern hardwood forest," Journal of Geophysical Research D, vol. 110, no. 7, Article ID D07301, pp. 1–12, 2005

149. H. Puxbaum and G. König, "Observation of dipropenyldisulfide and other organic sulfur compounds in the atmosphere of a beech forest with Allium ursinum ground cover," Atmospheric Environment, vol. 31, no. 2, pp. 291–294, 1997.

150. F. Rapparini, R. Baraldi, F. Miglietta, and F. Loreto, "Isoprenoid emission in trees of Quercus pubescensand Quercus ilex with

lifetime exposure to naturally high CO_2 environment," Plant, Cell and Environment, vol. 27, no. 4, pp. 381–391, 2004

151. R. A. Rasmussen, "Isoprene: identified as a forest-type emission to the atmosphere," Environmental Science and Technology, vol. 4, no. 8, pp. 667–671, 1970.

152. R. A. Rasmussen and F. Went, "Volatile organic material of plant origin in the atmosphere," Proceedings of the National Academy of Sciences, vol. 53, pp. 215–220, 1965.

153. R. C. Rhew, B. R. Miller, and R. F. Welss, "Natural methyl bromide and methyl chloride emissions from coastal salt marshes," Nature, vol. 403, no. 6767, pp. 292–295, 2000

154. J. M. Roberts, F. C. Fehsenfeld, D. L. Albritton, and R. E. Sievers, "Measurement of monoterpene hydrocarbons at Niwot Ridge, Colorado," Journal of Geophysical Research, vol. 88, no. 15, pp. 10.667–10.678, 1983.

155. J. M. Roberts, C. J. Hahn, F. C. Fehsenfeld, J. M. Warnock, D. L. Albritton, and R. E. Sievers, "Monoterpene hydrocarbons in the nighttime troposphere," Environmental Science and Technology, vol. 19, no. 4, pp. 364–369, 1985.

156. G. Sanadze, "The nature of gaseous substances emitted by leaves of Robinia pseudoacacia,"Soobshcheniya Akademi Nauk Gruzinskoj, vol. 27, pp. 747–750, 1957.

157. T. J. Savage, M. K. Hristova, and R. Croteau, "Evidence for an elongation/reduction/C1-elimination pathway in the biosynthesis of n-heptane in xylem of Jeffrey pine," Plant Physiology, vol. 111, no. 4, pp. 1263–1269, 1996.

158. S. Sawada and T. Totsuka, "Natural and anthropogenic sources and fate of atmospheric ethylene,"Atmospheric Environment, vol. 20, no. 5, pp. 821–832, 1986.

159. G. W. Schade and A. H. Goldstein, "Fluxes of oxygenated volatile organic compounds from a ponderosa pine plantation," Journal of Geophysical Research D, vol. 106, no. 3, pp. 3111–3123, 2001.

160. R. Seco, I. Filella, J. Llusià, and J. Peñuelas, "Methanol as a signal triggering isoprenoid emissions and photosynthetic performance in Quercus ilex," Acta Physiologiae Plantarum, vol. 33, no. 6, pp. 2413–2422, 2011

161. T. D. Sharkey, E. L. Singsaas, P. J. Vanderveer, and C. Geron, "Field measurements of isoprene emission from trees in response to temperature and light," Tree Physiology, vol. 16, no. 7, pp. 649–654, 1996.

162. T. D. Sharkey, E. L. Singsaas, M. T. Lerdau, and C. D. Geron, "Weather effects on isoprene emission capacity and applications in emissions algorithms," Ecological Applications, vol. 9, no. 4, pp. 1132–1137, 1999.

163. U. K. Sharma, Y. Kajii, and H. Akimoto, "Characterization of NMHCs in downtown urban center Kathmandu and rural site Nagarkot in Nepal," Atmospheric Environment, vol. 34, no. 20, pp. 3297–3307, 2000

164. R. W. Shaw Jr., A. L. Crittenden, R. K. Stevens, D. R. Cronn, and V. S. Titov, "Ambient concentrations of hydrocarbons from conifers in atmospheric gases and aerosol particles measured in Soviet Georgia,"Environmental Science and Technology, vol. 17, no. 7, pp. 389–395, 1983.

165. M. Šimpraga, H. Verbeeck, M. Demarcke et al., "Clear link between drought stress, photosynthesis and biogenic volatile organic compounds in Fagus sylvatica L," Atmospheric Environment, vol. 45, no. 30, pp. 5254–5259, 2011

166. B. C. Sive, R. K. Varner, H. Mao, D. R. Blake, O. W. Wingenter, and R. Talbot, "A large terrestrial source of methyl iodide," Geophysical Research Letters, vol. 34, no. 17, Article ID L17808, 2007. · ·

167. C. Spirig, A. Neftel, C. Ammann et al., "Eddy covariance flux measurements of biogenic VOCs during ECHO 2003 using proton transfer reaction mass spectrometry," Atmospheric Chemistry and Physics, vol. 5, no. 2, pp. 465–481, 2005.

168. M. Staudt, A. Ennajah, F. Mouillot, and R. Joffre, "Do volatile organic compound emissions of Tunisian cork oak populations originating from contrasting climatic conditions differ in their responses to summer drought?" Canadian Journal of Forest Research, vol. 38, no. 12, pp. 2965–2975, 2008. · ·

169. R. Steinbrecher, M. Klauer, K. Hauff et al., "Biogenic and anthropogenic fluxes of non-methane hydrocarbons over an urban-impacted forest, Frankfurter Stadtwald, Germany," Atmospheric Environment, vol. 34, no. 22, pp. 3779–3788, 2000.

170. A. Tani and Y. Kawawata, "Isoprene emission from the major native Quercus spp. in Japan,"Atmospheric Environment, vol. 42, no. 19, pp. 4540–4550, 2008. · ·

171. A. Tani, S. Nozoe, M. Aoki, and C. N. Hewitt, "Monoterpene fluxes measured above a Japanese red pine forest at Oshiba plateau, Japan," Atmospheric Environment, vol. 36, no. 21, pp. 3391–3402, 2002. · ·

172. D. T. Tingey, M. Manning, L. C. Grothaus, and W. F. Burns, "Influence of light and temperature on isoprene emission rates from live Oak," Physiologia Plantarum, vol. 47, no. 2, pp. 112–118, 1979.

173. D. T. Tingey, M. Manning, L. C. Grothaus, and W. F. Burns, "Influence of light and temperature on monoterpene emission rates from slash pine," Plant Physiology, vol. 65, no. 5, pp. 797–801, 1980.

174. J. K.-Y. Tsui, A. Guenther, W.-K. Yip, and F. Chen, "A biogenic volatile organic compound emission inventory for Hong Kong," Atmospheric Environment, vol. 43, no. 40, pp. 6442–6448, 2009. · ·

175. H. J. Wang, J. Y. Xia, Y. J. Mu, L. Nie, X. G. Han, and S. Q. Wan, "BVOCs emission in a semi-arid grassland under climate warming and nitrogen deposition," Atmospheric Chemistry and Physics, vol. 12, no. 8, pp. 3809–3819, 2012. · ·

176. C. Warneke, J. A. de Gouw, L. Del Negro et al., "Biogenic emission measurement and inventories determination of biogenic emissions in the eastern United States and Texas and comparison with biogenic emission inventories," Journal of Geophysical Research D, vol. 115, no. 5, Article ID D00F18, 2010. · ·

177. S. Welter, A. Bracho-Nunez, C. Mir et al., "The diversification of terpene emissions in Mediterranean oaks: lessons from a study of Quercus suber , Quercus canariensis and its hybrid Quercus afares," Tree Physiology, vol. 32, no. 9, pp. 1082–1091, 2012.

178. H. Westberg, B. Lamb, R. Hafer, A. Hills, P. Shepson, and C. Vogel, "Measurement of isoprene fluxes at the PROPHET site," Journal of Geophysical Research D, vol. 106, no. 20, pp. 24347–24358, 2001.

179. C. Wiedinmyer, S. Friedfeld, W. Baugh et al., "Measurement and analysis of atmospheric concentrations of isoprene and its

reaction products in central Texas," Atmospheric Environment, vol. 35, no. 6, pp. 1001–1013, 2001. · ·

180. C. Wiedinmyer, J. Greenberg, A. Guenther et al., "Ozarks Isoprene Experiment (OZIE): measurements and modeling of the 'isoprene volcano'," Journal of Geophysical Research D, vol. 110, no. 18, Article ID D18307, pp. 1–17, 2005.

181. A. J. Winters, M. A. Adams, T. M. Bleby et al., "Emissions of isoprene, monoterpene and short-chained carbonyl compounds from Eucalyptus spp. in southern Australia," Atmospheric Environment, vol. 43, no. 19, pp. 3035–3043, 2009. · ·

182. Z. Xiaoshan, M. Yujing, S. Wenzhi, and Z. Yahui, "Seasonal variations of isoprene emissions from deciduous trees," Atmospheric Environment, vol. 34, no. 18, pp. 3027–3032, 2000. · ·

183. A. Yani, G. Pauly, M. Faye, F. Salin, and M. Gleizes, "The effect of a long-term water stress on the metabolism and emission of terpenes of the foliage of Cupressus sempervirens," Plant, Cell and Environment, vol. 16, no. 8, pp. 975–981, 1993.

184. Y. Yokouchi, M. Okaniwa, Y. Ambe, and K. Fuwa, "Seasonal variation of monoterpenes in the atmosphere of a pine forest," Atmospheric Environment, vol. 17, no. 4, pp. 743–750, 1983.

185. Y. Yokouchi, A. Hijikata, and Y. Ambe, "Seasonal variation of monoterpene emission rate in a pine forest," Chemosphere, vol. 13, no. 2, pp. 255–259, 1984.

186. Y. Yokouchi and Y. Ambe, "Factors affecting the emission of monoterpenes from red pine (Pinus densiflora)," Plant Physiology, vol. 75, no. 4, pp. 1009–1012, 1984.

187. B. Baker, J.-H. Bai, C. Johnson et al., "Wet and dry season ecosystem level fluxes of isoprene and monoterpenes from a southeast Asian secondary forest and rubber tree plantation," Atmospheric Environment, vol. 39, no. 2, pp. 381–390, 2005. · ·

188. D. R. Cronn and W. Nutmagul, "Analysis of atmospheric hydrocarbons during winter MONEX (Borneo)," Tellus, vol. 34, no. 2, pp. 159–165, 1982.

189. P. Crutzen, M. Coffey, A. Delany et al., "Observations of air composition in Brazil between the equator and 20°S during the dry season," Acta Amazonica, vol. 15, pp. 77–119, 1985.

190. L. Donoso, R. Romero, A. Rondón, E. Fernandez, P. Oyola, and E. Sanhueza, "Natural and anthropogenic C_2 to C_6 hydrocarbons in the Central-Eastern Venezuelan atmosphere during the rainy season," Journal of Atmospheric Chemistry, vol. 25, no. 2, pp. 201–214, 1996.

191. C. Geron, A. Guenther, J. Greenberg, H. W. Loescher, D. Clark, and B. Baker, "Biogenic volatile organic compound emissions from a lowland tropical wet forest in Costa Rica," Atmospheric Environment, vol. 36, no. 23, pp. 3793–3802, 2002.

192. J. P. Greenberg, P. R. Zimmerman, L. Heidt, and W. Pollock, "Hydrocarbon and carbon monoxide emissions from biomass burning in Brazil," Journal of Geophysical Research, vol. 89, no. 1, pp. 1350–1354, 1984.

193. J. P. Greenberg, "Biogenic volatile organic compound emissions in central Africa during the Experiment for the Regional Sources and Sinks of Oxidants (EXPRESSO) biomass burning season," Journal of Geophysical Research D, vol. 104, no. 23, pp. 30659–30671, 1999.

194. J. P. Greenberg, A. Guenther, P. Harley et al., "Eddy flux and leaf-level measurements of biogenic VOC emissions from mopane woodland of Botswana," Journal of Geophysical Research D, vol. 108, no. 13, pp. 2–9, 2003.

195. J. P. Greenberg, A. B. Guenther, G. Pétron et al., "Biogenic VOC emissions from forested Amazonian landscapes," Global Change Biology, vol. 10, no. 5, pp. 651–662, 2004. · ·

196. G. Gregory, R. Harriss, R. Talbot et al., "Air chemistry over the tropical forest of Guyana," Journal of Geophysical Research, vol. 91, pp. 8603–8612, 1986.

197. A. Guenther, L. Otter, P. Zimmerman, J. Greenberg, R. Scholes, and M. Scholes, "Biogenic hydrocarbon emissions from southern African savannas," Journal of Geophysical Research D, vol. 101, no. 20, pp. 25859–25865, 1996.

198. P. Harley, P. Vasconcellos, L. Vierling et al., "Variation in potential for isoprene emissions among Neotropical forest sites," Global Change Biology, vol. 10, no. 5, pp. 630–650, 2004.

199. D. Helmig, B. Balsley, K. Davis et al., "Vertical profiling and determination of landscape fluxes of biogenic nonmethane

hydrocarbons within the planetary boundary layer in the Peruvian Amazon,"Journal of Geophysical Research D, vol. 103, no. 19, pp. 25519–25532, 1998.

200. R. Holzinger, E. Sanhueza, R. von Kuhlmann, B. Kleiss, L. Donoso, and P. J. Crutzen, "Diurnal cycles and seasonal variation of isoprene and its oxidation products in the tropical savanna atmosphere,"Global Biogeochemical Cycles, vol. 16, no. 4, pp. 22–1, 2002.

201. T. Karl, M. Potosnak, A. Guenther et al., "Exchange processes of volatile organic compounds above a tropical rain forest: implications for modeling tropospheric chemistry above dense vegetation," Journal of Geophysical Research D, vol. 109, no. 18, pp. D18306–19, 2004.

202. T. Karl, A. Guenther, R. J. Yokelson et al., "The tropical forest and fire emissions experiment: emission, chemistry, and transport of biogenic volatile organic compounds in the lower atmosphere over Amazonia," Journal of Geophysical Research D, vol. 112, no. 18, Article ID D18302, 2007. · ·

203. T. Karl, A. Guenther, A. Turnipseed, G. Tyndall, P. Artaxo, and S. Martin, "Rapid formation of isoprene photo-oxidation products observed in Amazonia," Atmospheric Chemistry and Physics, vol. 9, no. 20, pp. 7753–7767, 2009.

204. M. Keller and M. Lerdau, "Isoprene emission from tropical forest canopy leaves," Global Biogeochemical Cycles, vol. 13, no. 1, pp. 19–29, 1999. ·

205. J. Kesselmeier, U. Kuhn, A. Wolf et al., "Atmospheric volatile organic compounds (VOC) at a remote tropical forest site in central Amazonia," Atmospheric Environment, vol. 34, no. 24, pp. 4063–4072, 2000.

206. J. Kesselmeier, U. Kuhn, S. Rottenberger et al., "Concentrations and species composition of atmospheric volatile organic compounds (VOCs) as observed during the wet and dry season in Rondônia (Amazonia)," Journal of Geophysical Research D, vol. 107, no. 20, pp. 1–20, 2002. · ·

207. L. F. Klinger, "Patterns in volatile organic compound emissions along a savanna-rainforest gradient in central Africa," Journal of Geophysical Research D, vol. 103, no. 1, pp. 1443–1454, 1998.

208. U. Kuhn, S. Rottenberger, T. Biesenthal et al., "Isoprene and monoterpene emissions of Amazonian tree species during the wet season: direct and indirect investigations on controlling environmental functions," Journal of Geophysical Research D, vol. 107, no. 20, pp. XCXLIII–XCXLIV, 2002

209. U. Kuhn, S. Rottenberger, T. Biesenthal et al., "Seasonal differences in isoprene and light-dependent monoterpene emission by Amazonian tree species," Global Change Biology, vol. 10, no. 5, pp. 663–682, 2004.

210. U. Kuhn, M. O. Andreae, C. Ammann et al., "Isoprene and monoterpene fluxes from Central Amazonian rainforest inferred from tower-based and airborne measurements, and implications on the atmospheric chemistry and the local carbon budget," Atmospheric Chemistry and Physics, vol. 7, no. 11, pp. 2855–2879, 2007.

211. C. E. Jones, J. R. Hopkins, and A. C. Lewis, "In situ measurements of isoprene and monoterpenes within a south-east Asian tropical rainforest," Atmospheric Chemistry and Physics, vol. 11, no. 14, pp. 6971–6984, 2011

212. B. Langford, P. K. Misztal, E. Nemitz et al., "Fluxes and concentrations of volatile organic compounds from a South-East Asian tropical rainforest," Atmospheric Chemistry and Physics, vol. 10, no. 17, pp. 8391–8412, 2010.

213. D. Y. C. Leung, P. Wong, B. K. H. Cheung, and A. Guenther, "Improved land cover and emission factors for modeling biogenic volatile organic compounds emissions from Hong Kong," Atmospheric Environment, vol. 44, no. 11, pp. 1456–1468, 2010.

214. P. K. Misztal, S. M. Owen, A. B. Guenther et al., "Large estragole fluxes from oil palms in Borneo,"Atmospheric Chemistry and Physics, vol. 10, no. 9, pp. 4343–4358, 2010.

215. P. K. Misztal, E. Nemitz, B. Langford et al., "Direct ecosystem fluxes of volatile organic compounds from oil palms in South-East Asia," Atmospheric Chemistry and Physics, vol. 11, no. 17, pp. 8995–9017, 2011.

216. J.-F. Müller, T. Stavrakou, S. Wallens et al., "Global isoprene emissions estimated using MEGAN, ECMWF analyses and a detailed canopy environment model," Atmospheric Chemistry and Physics, vol. 8, no. 5, pp. 1329–1341, 2008.

217. H. Oku, M. Fukuta, H. Iwasaki, P. Tambunan, and S. Baba, "Modification of the isoprene emission model G93 for tropical tree Ficus virgata," Atmospheric Environment, vol. 42, no. 38, pp. 8747–8754, 2008.

218. P. K. Padhy and C. K. Varshney, "Isoprene emission from tropical tree species," Environmental Pollution, vol. 135, no. 1, pp. 101–109, 2005

219. R. A. Rasmussen and M. A. K. Khalil, "Isoprene over the Amazon Basin," Journal of Geophysical Research, vol. 93, no. 2, pp. 1417–1421, 1988.

220. H. J. I. Rinne, A. B. Guenther, J. P. Greenberg, and P. C. Harley, "Isoprene and monoterpene fluxes measured above Amazonian rainforest and their dependence on light and temperature," Atmospheric Environment, vol. 36, no. 14, pp. 2421–2426, 2002.

221. T. Saito, Y. Yokouchi, Y. Kosugi, M. Tani, E. Philip, and T. Okuda, "Methyl chloride and isoprene emissions from tropical rain forest in Southeast Asia," Geophysical Research Letters, vol. 35, no. 19, Article ID L19812, 2008.

222. E. Sanhueza, M. Santana, D. Trapp et al., "Field measurement evidence for an atmospheric chemical source of formic and acetic acids in the tropic," Geophysical Research Letters, vol. 23, no. 9, pp. 1045–1048, 1996.

223. J. E. Saxton, A. C. Lewis, J. H. Kettlewell et al., "Isoprene and monoterpene measurements in a secondary forest in northern Benin," Atmospheric Chemistry and Physics, vol. 7, no. 15, pp. 4095–4106, 2007.

224. D. Sercanda, A. Guenther, L. Klinger et al., "EXPRESSO flux measurements at upland and lowland Congo tropical forest site," Tellus B, vol. 53, no. 3, pp. 220–234, 2001.

225. R. W. Talbot, M. O. Andreae, H. Berresheim, D. J. Jacob, and K. M. Beecher, "Sources and sinks of formic, acetic, and pyruvic acids over central Amazonia. 2. Wet season," Journal of Geophysical Research, vol. 95, no. 10, pp. 16–811, 1990.

226. C. K. Varshney and A. P. Singh, "Isoprene emission from Indian trees," Journal of Geophysical Research D, vol. 108, no. 24, pp. 24–7, 2003.

227. C. Warneke, S. L. Luxembourg, J. A. de Gouw, H. J. I. Rinne, A. B. Guenther, and R. Fall, "Disjunct eddy covariance measurements of oxygenated volatile organic compounds fluxes from an alfalfa field before and after cutting," Journal of Geophysical Research D, vol. 107, no. 7-8, pp. 6–1, 2002.

228. J. Williams, U. Pöschl, P. J. Crutzen et al., "An atmospheric chemistry interpretation of mass scans obtained from a proton transfer mass spectrometer flown over the tropical rainforest of Surinam,"Journal of Atmospheric Chemistry, vol. 38, no. 2, pp. 133–166, 2001. · ·

229. P. R. Zimmerman, J. P. Greenberg, and C. E. Westberg, "Measurements of atmospheric hydrocarbons and biogenic emission fluxes in the Amazon Boundary Layer," Journal of Geophysical Research, vol. 93, no. 2, pp. 1407–1416, 1988.

230. J. Bäck, J. Aalto, M. Henriksson, H. Hakola, Q. He, and M. Boy, "Chemodiversity in terpene emissions at a boreal Scots pine stand," Biogeosciences Discussions, vol. 8, no. 5, pp. 10577–10615, 2011. · ·

231. J. Bai, F. Lin, X. Wan, A. Guenther, A. Turnipseed, and T. Duhl, "Volatile organic compound emission fluxes from a temperate forest in Changbai Mountain," Acta Scientiae Circumstantiae, vol. 32, no. 3, pp. 545–554, 2012.

232. A. Ekberg, A. Arneth, H. Hakola, S. Hayward, and T. Holst, "Isoprene emission from wetland sedges,"Biogeosciences, vol. 6, no. 4, pp. 601–613, 2009.

233. A. Ekberg, A. Arneth, and T. Holst, "Isoprene emission from Sphagnum species occupying different growth positions above the water table," Boreal Environment Research, vol. 16, no. 1, pp. 47–59, 2011.

234. P. Faubert, P. Tiiva, Å. Rinnan, A. Michelsen, J. K. Holopainen, and R. Rinnan, "Doubled volatile organic compound emissions from subarctic tundra under simulated climate warming," New Phytologist, vol. 187, no. 1, pp. 199–208, 2010. · ·

235. I. Filella, M. J. Wilkinson, J. Llusià, C. N. Hewitt, and J. Peñuelas, "Volatile organic compounds emissions in Norway spruce (Picea abies) in response to temperature changes," Physiologia Plantarum, vol. 130, no. 1, pp. 58–66, 2007.

236. J. D. Fuentes, D. Wang, and L. Gu, "Seasonal variations in isoprene emissions from a boreal aspen forest," Journal of Applied Meteorology, vol. 38, no. 7, pp. 855–869, 1999.

237. A. Ghirardo, K. Koch, R. Taipale, I. Zimmer, J.-P. Schnitzler, and J. Rinne, "Determination of de novo and pool emissions of terpenes from four common boreal/alpine trees by 13CO$_2$ labelling and PTR-MS analysis," Plant, Cell and Environment, vol. 33, no. 5, pp. 781–792, 2010. · ·

238. S. Haapanala, J. Rinne, K.-H. Pystynen, H. Hellén, H. Hakola, and T. Riutta, "Measurements of hydrocarbon emissions from a boreal fen using the REA technique," Biogeosciences, vol. 3, no. 1, pp. 103–112, 2006.

239. H. Hakola, J. Rinne, and T. Laurila, "The hydrocarbon emission rates of tea-leafed willow (Salix phylicifolia), silver birch (Betula pendula) and European aspen (Populus tremula)," Atmospheric Environment, vol. 32, no. 10, pp. 1825–1833, 1998. · ·

240. H. Hakola, T. Laurila, J. Rinne, and K. Puhto, "The ambient concentrations of biogenic hydrocarbons at a northern European, boreal site," Atmospheric Environment, vol. 34, no. 29-30, pp. 4971–4982, 2000.

241. H. Hakola, T. Laurila, V. Lindfors, H. Hellén, A. Gaman, and J. Rinne, "Variation of the VOC emission rates of birch species during the growing season," Boreal Environment Research, vol. 6, no. 3, pp. 237–249, 2001.

242. H. Hakola, V. Tarvainen, J. Bäck et al., "Seasonal variation of mono- and sesquiterpene emission rates of Scots pine," Biogeosciences, vol. 3, no. 1, pp. 93–101, 2006.

243. D. T. Hanson, S. Swanson, L. E. Graham, and T. D. Sharkey, "Evolutionary significance of isoprene emission from mosses," The American Journal of Botany, vol. 86, no. 5, pp. 634–639, 1999.

244. H. Hellén, H. Hakola, K.-H. Pystynen, J. Rinne, and S. Haapanala, "C$_2$-C$_{10}$ hydrocarbon emissions from a boreal wetland and forest floor," Biogeosciences, vol. 3, no. 2, pp. 167–174, 2006.

245. T. Holst, A. Arneth, S. Hayward et al., "BVOC ecosystem flux measurements at a high latitude wetland site," Atmospheric Chemistry and Physics, vol. 10, no. 4, pp. 1617–1634, 2010.

246. O. Hov, J. Schjoldager, and B. M. Wathne, "Measurement and modeling of the concentrations of terpenes in coniferous forest air (Norway)," Journal of Geophysical Research, vol. 88, no. 15, pp. 10679–10688, 1983.

247. R. Janson, "Monoterpene concentrations in and above a forest of Scots pine," Journal of Atmospheric Chemistry, vol. 14, no. 1–4, pp. 385–394, 1992. ·

248. R. Janson and C. de Serves, "Isoprene emissions from boreal wetlands in Scandinavia," Journal of Geophysical Research D, vol. 103, no. 19, pp. 25513–25517, 1998.

249. R. Janson, C. de Serves, and R. Romero, "Emission of isoprene and carbonyl compounds from a boreal forest and wetland in Sweden," Agricultural and Forest Meteorology, vol. 98-99, pp. 671–681, 1999. · ·

250. B. T. Jobson, Z. Wu, H. Niki, and L. A. Barrie, "Seasonal trends of isoprene, C_2-C_5 alkanes, and acetylene at a remote boreal site in Canada," Journal of Geophysical Research, vol. 99, pp. 1589–1599, 1994.

251. K. Kempf, E. Allwine, H. Westberg, C. Claiborn, and B. Lamb, "Hydrocarbon emissions from spruce species using environmental chamber and branch enclosure methods," Atmospheric Environment, vol. 30, no. 9, pp. 1381–1389, 1996. · ·

252. L. F. Klinger, P. R. Zimmerman, J. P. Greenberg, L. E. Heidt, and A. B. Guenther, "Carbon trace gas fluxes along a successional gradient in the Hudson-Bay Lowland," Journal of Geophysical Research, vol. 99, no. 1, pp. 1469–1494, 1994.

253. D. M. Martin, J. Gershenzon, and J. Bohlmann, "Induction of volatile terpene biosynthesis and diurnal emission by methyl jasmonate in foliage of Norway spruce," Plant Physiology, vol. 132, no. 3, pp. 1586–1599, 2003.

254. E. Pattey, R. L. Desjardins, H. Westberg, B. Lamb, and T. Zhu, "Measurement of isoprene emissions over a black spruce stand using a tower-based relaxed eddy-accumulation system," Journal of Applied Meteorology, vol. 38, no. 7, pp. 870–877, 1999.

255. G. Petersson, "High ambient concentrations of monoterpenes in a Scandinavian pine forest,"Atmospheric Environment, vol. 22, no. 11, pp. 2617–2619, 1988.

256. M. J. Potosnak, B. Baker, L. LeStourgeon et al., "Isoprene emissions from a tundra ecosystem," Biogeosciences, vol. 10, pp. 871–889, 2013.

257. T. Räisänen, A. Ryyppö, and S. Kellomäki, "Monoterpene emission of a boreal Scots pine (Pinus sylvestris L.) forest," Agricultural and Forest Meteorology, vol. 149, no. 5, pp. 808–819, 2009.

258. J. Rinne, H. Hakola, and T. Laurila, "Vertical fluxes of monoterpenes above a Scots pine stand in the boreal vegetation zone," Physics and Chemistry of the Earth B, vol. 24, no. 6, pp. 711–715, 1999. · ·

259. J. Rinne, H. Hakola, T. Laurila, and Ü. Rannik, "Canopy scale monoterpene emissions of Pinus sylvestrisdominated forests," Atmospheric Environment, vol. 34, no. 7, pp. 1099–1107, 2000. · ·

260. J. Rinne, R. Taipale, T. Markkanen et al., "Hydrocarbon fluxes above a Scots pine forest canopy: measurements and modeling," Atmospheric Chemistry and Physics, vol. 7, no. 12, pp. 3361–3372, 2007.

261. T. M. Ruuskanen, H. Hakola, M. K. Kajos, H. Hellén, V. Tarvainen, and J. Rinne, "Volatile organic compound emissions from Siberian larch," Atmospheric Environment, vol. 41, no. 27, pp. 5807–5812, 2007. · ·

262. C. Spirig, A. Guenther, J. P. Greenberg, P. Calanca, and V. Tarvainen, "Tethered balloon measurements of biogenic volatile organic compounds at a Boreal forest site," Atmospheric Chemistry and Physics, vol. 4, no. 1, pp. 215–229, 2004.

263. V. Tarvainen, H. Hakola, H. Hellén, J. Bäck, P. Hari, and M. Kulmala, "Temperature and light dependence of the VOC emissions of Scots pine," Atmospheric Chemistry and Physics, vol. 5, no. 4, pp. 989–998, 2005.

264. P. Tiiva, R. Rinnan, T. Holopainen, S. K. Mörsky, and J. K. Holopainen, "Isoprene emissions from boreal peatland microcosms; effects of elevated ozone concentration in an open field experiment," Atmospheric Environment, vol. 41, no. 18, pp. 3819–3828, 2007. · ·

265. P. Tiiva, P. Faubert, A. Michelsen, T. Holopainen, J. K. Holopainen, and R. Rinnan, "Climatic warming increases isoprene emission

from a subarctic heath," New Phytologist, vol. 180, no. 4, pp. 853–863, 2008. · ·

266. T. Vuorinen, A.-M. Nerg, E. Vapaavuori, and J. K. Holopainen, "Emission of volatile organic compounds from two silver birch (Betula pendula Roth) clones grown under ambient and elevated CO_2 and different O_3 concentrations," Atmospheric Environment, vol. 39, no. 7, pp. 1185–1197, 2005. · ·

267. Q.-H. Zhang, F. Schlyter, and P. Anderson, "Green leaf volatiles interrupt pheromone response of spruce bark beetle, Ips typographus," Journal of Chemical Ecology, vol. 25, no. 12, pp. 2847–2861, 1999.

268. T. Zhu, D. Wang, R. L. Desjardins, and J. I. Macpherson, "Aircraft-based volatile organic compounds flux measurements with relaxed eddy accumulation," Atmospheric Environment, vol. 33, no. 12, pp. 1969–1979, 1999. · ·

269. N. G. Agelopoulos, K. Chamberlain, and J. A. Pickett, "Factors affecting volatile emissions of intact potato plants, Solanum tuberosum: variability of quantities and stability of ratios," Journal of Chemical Ecology, vol. 26, no. 2, pp. 497–511, 2000. · ·

270. J. Arey, "Terpenes emitted from agricultural species found in California›s Central Valley," Journal of Geophysical Research, vol. 96, no. 5, pp. 9329–9336, 1991.

271. J. Arey, A. M. Winer, R. Atkinson, S. M. Aschmann, W. D. Long, and C. L. Morrison, "The emission of (Z)-3-hexen-1-ol, (Z)-3-hexenylacetate and other oxygenated hydrocarbons from agricultural plant species," Atmospheric Environment, vol. 25, no. 5-6, pp. 1063–1075, 1991.

272. J. Arey, D. E. Crowley, M. Crowley, M. Resketo, and J. Lester, "Hydrocarbon emissions from natural vegetation in California›s South Coast Air Basin," Atmospheric Environment, vol. 29, no. 21, pp. 2977–2988, 1995.

273. N. Copeland, J. N. Cape, and M. R. Heal, "Volatile organic compound emissions from Miscanthus and short rotation coppice willow bioenergy crops," Atmospheric Environment, vol. 60, pp. 327–335, 2012.

274. C. M. de Moraes, M. C. Mescher, and J. H. Tumlinson, "Caterpillar-induced nocturnal plant volatiles repel conspecific females," Nature, vol. 410, no. 6828, pp. 577–579, 2001. · ·

275. A. S. D. Eller, K. Sekimoto, J. B. Gilman et al., "Volatile organic compound emissions from switchgrass cultivars used as biofuel crops," Atmospheric Environment, vol. 45, no. 19, pp. 3333–3337, 2011. · ·

276. S. P. Gouinguené and T. C. J. Turlings, "The effects of abiotic factors on induced volatile emissions in corn plants," Plant Physiology, vol. 129, no. 3, pp. 1296–1307, 2002. · ·

277. C. N. Hewitt, R. K. Monson, and R. Fall, "Isoprene emissions from the grass Arundo donax L. are not linked to photorespiration," Plant Science, vol. 66, no. 2, pp. 139–144, 1990.

278. F. Loreto and T. D. Sharkey, "Isoprene emission by plants is affected by transmissible wound signals,"Plant Cell and Environment, vol. 16, no. 5, pp. 563–570, 1993.

279. J. Ruther and S. Kleier, "Plant-plant signaling: ethylene synergizes volatile emission in Zea mays induced by exposure to (Z)-3-hexen-1-ol," Journal of Chemical Ecology, vol. 31, no. 9, pp. 2217–2222, 2005

280. G. Schuh, A. C. Heiden, T. Hoffmann et al., "Emissions of volatile organic compounds from sunflower and beech: dependence on temperature and light intensity," Journal of Atmospheric Chemistry, vol. 27, no. 3, pp. 291–318, 1997.

281. A. Tava, N. Berardo, C. Cunico, M. Romani, and M. Odoardi, "Cultivar differences and seasonal changes of primary metabolites and flavor constituents in tall fescue in relation to palatability," Journal of Agricultural and Food Chemistry, vol. 43, no. 1, pp. 98–101, 1995.

282. C. Warneke, S. L. Luxembourg, J. A. de Gouw, H. J. I. Rinne, A. B. Guenther, and R. Fall, "Disjunct eddy covariance measurements of oxygenated volatile organic compounds fluxes from an alfalfa field before and after cutting," Journal of Geophysical Research D, vol. 107, no. 7-8, pp. 6–1, 2002.

283. S. Juuti, J. Arey, and R. Atkinson, "Monoterpene emission rate measurements from a monterey pine,"Journal of Geophysical Research, vol. 95, no. 6, pp. 7515–7519, 1990.

284. A. Guenther, P. Zimmerman, and M. Wildermuth, "Natural volatile organic compound emission rate estimates for U.S. woodland landscapes," Atmospheric Environment, vol. 28, no. 6, pp. 1197–1210, 1994.

285. S. Kim, A. Guenther, and E. Apel, "Quantitative and qualitative sensing techniques for biogenic volatile organic compounds and their oxidation products," Environmental Science, 2013.

286. T. Karl, E. Apell, A. Hodzic, D. D. Riemer, D. R. Blake, and C. Wiedinmyer, "Emissions of volatile organic compounds inferred from airborne flux measurements over a megacity," Atmospheric Chemistry and Physics, vol. 9, no. 1, pp. 271–285, 2009.

287. A. Guenther, C. Geron, T. Pierce, B. Lamb, P. Harley, and R. Fall, "Natural emissions of non-methane volatile organic compounds, carbon monoxide, and oxides of nitrogen from North America,"Atmospheric Environment, vol. 34, no. 12–14, pp. 2205–2230, 2000.

288. J. Kesselmeier, A. Guenther, T. Hoffmann, M. T. Piedade, and J. Warnke, "Natural volatile organic compound emissions from plants and their roles in oxidant balance and particle formation," inAmazonia and Global Change, M. Keller, Ed., Geophysical Monograph Series, 2009.

289. A. Guenther, M. Kulmala, A. Turnipseed, J. Rinne, T. Suni, and A. Reissell, "Integrated land ecosystem-atmosphere processes study (iLEAPS) assessment of global observational networks," Boreal Environment Research, vol. 16, no. 4, pp. 321–336, 2011.

290. A. P. Altshuller, "Review: natural volatile organic substances and their effect on air quality in the United States," Atmospheric Environment, vol. 17, no. 11, pp. 2131–2165, 1983.

291. M. Trainer, "Models and observations of the impact of natural hydrocarbons on rural ozone," Nature, vol. 329, no. 6141, pp. 705–707, 1987.

292. W. L. Chameides, R. W. Lindsay, J. Richardson, and C. S. Kiang, "The role of biogenic hydrocarbons in urban photochemical smog: atlanta as a case study," Science, vol. 241, no. 4872, pp. 1473–1475, 1988.

293. C. R. Hoyle, M. Boy, N. M. Donahue et al., "A review of the anthropogenic influence on biogenic secondary organic aerosol," Atmospheric Chemistry and Physics, vol. 11, no. 1, pp. 321–343, 2011. · ·

294. J. T. Knudsen, R. Eriksson, J. Gershenzon, and B. Ståhl, "Diversity and distribution of floral scent,"Botanical Review, vol. 72, no. 1, pp. 1–120, 2006. · ·

295. J. H. Langenheim, "Higher plant terpenoids: a phytocentric overview of their ecological roles," Journal of Chemical Ecology, vol. 20, no. 6, pp. 1223–1280, 1994. · ·

296. T. R. Duhl, D. Helmig, and A. Guenther, "Sesquiterpene emissions from vegetation: a review,"Biogeosciences, vol. 5, no. 3, pp. 761–777, 2008.

297. N. C. Bouvier-Brown, A. H. Goldstein, J. B. Gilman, W. C. Kuster, and J. A. de Gouw, "In-situ ambient quantification of monoterpenes, sesquiterpenes and related oxygenated compounds during BEARPEX 2007: implications for gas- and particle-phase chemistry," Atmospheric Chemistry and Physics, vol. 9, no. 15, pp. 5505–5518, 2009.

298. T. Sakulyanontvittaya, A. Guenther, D. Helmig, J. Milford, and C. Wiedinmyer, "Secondary organic aerosol from sesquiterpene and monoterpene emissions in the United States," Environmental Science and Technology, vol. 42, no. 23, pp. 8784–8790, 2008. · ·

299. D. W. Gray, M. T. Lerdau, and A. H. Goldstein, "Influences of temperature history, water stress, and needle age on methylbutenol emissions," Ecology, vol. 84, no. 3, pp. 765–776, 2003.

300. D. W. Gray, S. R. Breneman, L. A. Topper, and T. D. Sharkey, "Biochemical characterization and homology modeling of methylbutenol synthase and implications for understanding hemiterpene synthase evolution in plants," Journal of Biological Chemistry, vol. 286, no. 23, pp. 20582–20590, 2011. · ·

301. G.-I. Arimura, K. Matsui, and J. Takabayashi, "Chemical and molecular ecology of herbivore-induced plant volatiles: proximate factors and their ultimate functions," Plant and Cell Physiology, vol. 50, no. 5, pp. 911–923, 2009. · ·

302. J. R. Snider and G. A. Dawson, "Tropospheric light alcohols, carbonyls, and acetonitrile: concentrations in the southwestern United States and Henry's law data," Journal of Geophysical Research, vol. 90, pp. 3797–3805, 1985.

303. C. Warneke, T. Karl, H. Judmaier et al., "Acetone, methanol, and other partially oxidized volatile organic emissions from dead plant matter by abiological processes: significance for atmospheric HO(X) chemistry," Global Biogeochemical Cycles, vol. 13, no. 1, pp. 9–17, 1999. · ·

304. D. J. Jacob, B. D. Field, Q. Li et al., "Global budget of methanol: constraints from atmospheric observations," Journal of Geophysical Research D, vol. 110, no. 8, pp. 1–17, 2005. · ·

305. D. B. Millet, D. J. Jacob, T. G. Custer et al., "New constraints on terrestrial and oceanic sources of atmospheric methanol," Atmospheric Chemistry and Physics, vol. 8, no. 23, pp. 6887–6905, 2008.

306. T. Stavrakou, A. Guenther, A. Razavi et al., "First space-based derivation of the global atmospheric methanol emission fluxes," Atmospheric Chemistry and Physics, vol. 11, no. 10, pp. 4873–4898, 2011. · ·

307. D. J. Jacob, B. D. Field, E. M. Jin et al., "Atmospheric budget of acetone," Journal of Geophysical Research D, vol. 107, no. 9-10, pp. 5–1, 2002.

308. E. V. Fischer, D. J. Jacob, D. B. Millet, R. M. Yantosca, and J. Mao, "The role of the ocean in the global atmospheric budget of acetone," Geophysical Research Letters, vol. 39, no. 1, Article ID L01807, 2012. · ·

309. J. Kesselmeier, "Exchange of short-chain oxygenated volatile organic compounds (VOCs) between plants and the atmosphere: a compilation of field and laboratory studies," Journal of Atmospheric Chemistry, vol. 39, no. 3, pp. 219–233, 2001. · ·

310. D. B. Millet, A. Guenther, D. Siegel et al., "Global atmospheric budget of acetaldehyde: 3-D model analysis and constraints from in-situ and satellite observations," Atmospheric Chemistry and Physics, vol. 10, no. 7, pp. 3405–3425, 2010.

311. T. Stavrakou, J.-F. Müller, J. Peeters et al., "Satellite evidence for a large source of formic acid from boreal and tropical forests," Nature Geoscience, vol. 5, no. 1, pp. 26–30, 2012. · ·

312. J. Engelberth, H. T. Alborn, E. A. Schmelz, and J. H. Tumlinson, "Airborne signals prime plants against insect herbivore attack," Proceedings of the National Academy of Sciences of the United States of America, vol. 101, no. 6, pp. 1781–1785, 2004. · ·

313. T. C. Turlings and J. Ton, "Exploiting scents of distress: the prospect of manipulating herbivore-induced plant odours to enhance the control of agricultural pests," Current Opinion in Plant Biology, vol. 9, no. 4, pp. 421–427, 2006. · ·

314. C. C. Von Dahl, M. Hävecker, R. Schlögl, and I. T. Baldwin, "Caterpillar-elicited methanol emission: a new signal in plant-herbivore interactions?" Plant Journal, vol. 46, no. 6, pp. 948–960, 2006. · ·

315. K. Hüve, M. M. Christ, E. Kleist et al., "Simultaneous growth and emission measurements demonstrate an interactive control of methanol release by leaf expansion and stomata," Journal of Experimental Botany, vol. 58, no. 7, pp. 1783–1793, 2007. · ·

316. M. R. Kant, P. M. Bleeker, M. V. Wijk, R. C. Schuurink, and M. A. Haring, "Plant volatiles in defence,"Advances in Botanical Research, vol. 51, pp. 613–666, 2009. · ·

317. F. A. M. Wellburn and A. R. Wellburn, "Variable patterns of antioxidant protection but similar ethene emission differences in several ozone-sensitive and ozone-tolerant plant selections," Plant, Cell and Environment, vol. 19, no. 6, pp. 754–760, 1996.

318. J. Browse and G. A. Howe, "New weapons and a rapid response against insect attack," Plant physiology, vol. 146, no. 3, pp. 832–838, 2008. · ·

319. A. Hansjakob, M. Riederer, and U. Hildebrandt, "Wax matters: absence of very-long-chain aldehydes from the leaf cuticular wax of the glossy11 mutant of maize compromises the prepenetration processes of Blumeria graminis," Plant Pathology, vol. 60, no. 6, pp. 1151–1161, 2011. · ·

320. D. Chachalis, K. N. Reddy, and C. D. Elmore, "Characterization of leaf surface, wax composition, and control of redvine and trumpetcreeper with glyphosate," Weed Science, vol. 49, no. 2, pp. 156–163, 2001.

321. T. Karl, P. Harley, A. Guenther et al., "The bi-directional exchange of oxygenated VOCs between a loblolly pine (Pinus taeda) plantation and the atmosphere," Atmospheric Chemistry and Physics, vol. 5, no. 11, pp. 3015–3031, 2005.

322. M. A. H. Khan, M. E. Whelan, and R. C. Rhew, "Effects of temperature and soil moisture on methyl halide and chloroform fluxes from drained peatland pasture soils," Journal of Environmental Monitoring, vol. 14, no. 1, pp. 241–249, 2012. · ·

323. Y. Yoshida, Y. Wang, C. Shim, D. Cunnold, D. R. Blake, and G. S. Dutton, "Inverse modeling of the global methyl chloride sources,"

Journal of Geophysical Research D, vol. 111, no. 16, Article ID D16307, 2006. · ·

324. R. C. Rhew, "Sources and sinks of methyl bromide and methyl chloride in the tallgrass prairie: applying a stable isotope tracer technique over highly variable gross fluxes," Journal of Geophysical Research G, vol. 116, no. 3, Article ID G03026, 2011. · ·

325. T. S. Bates, B. K. Lamb, A. Guenther, J. Dignon, and R. E. Stoiber, "Sulfur emissions to the atmosphere from natural sources," Journal of Atmospheric Chemistry, vol. 14, no. 1–4, pp. 315–337, 1992. · ·

326. S. F. Watts, "The mass budgets of carbonyl sulfide, dimethyl sulfide, carbon disulfide and hydrogen sulfide," Atmospheric Environment, vol. 34, no. 5, pp. 761–779, 2000. · ·

327. J. Wildt, K. Kobel, G. Schuh-Thomas, and A. C. Heiden, "Emissions of oxygenated volatile organic compounds from plants part II: emissions of saturated aldehydes," Journal of Atmospheric Chemistry, vol. 45, no. 2, pp. 173–196, 2003.

328. F. Keppler, J. T. G. Hamilton, M. Braß, and T. Röckmann, "Methane emissions from terrestrial plants under aerobic conditions," Nature, vol. 439, no. 7073, pp. 187–191, 2006. · ·

329. T. A. Dueck, R. de Visser, H. Poorter et al., "No evidence for substantial aerobic methane emission by terrestrial plants: a 13C-labelling approach," New Phytologist, vol. 175, no. 1, pp. 29–35, 2007. · ·

330. S.-L. Steenhuisen, R. A. Raguso, A. Jürgens, and S. D. Johnson, "Variation in scent emission among floral parts and inflorescence developmental stages in beetle-pollinated Protea species (Proteaceae)," South African Journal of Botany, vol. 76, no. 4, pp. 779–787, 2010. · ·

331. T. E. Pierce and P. S. Waldruff, "PC-BEIS: a personal computer version of the Biogenic Emissions Inventory System," Journal of the Air and Waste Management Association, vol. 41, no. 7, pp. 937–941, 1991.

332. M. T. Benjamin and A. M. Winer, "Estimating the ozone-forming potential of urban trees and shrubs," Atmospheric Environment, vol. 32, no. 1, pp. 53–68, 1998.

333. P. C. Harley, R. K. Monson, and M. T. Lerdau, "Ecological and evolutionary aspects of isoprene emission from plants," Oecologia, vol. 118, no. 2, pp. 109–123, 1999. · ·

334. R. G. Latta, Y. B. Linhart, M. A. Snyder, and L. Lundquist, "Patterns of variation and correlation in the monoterpene composition of xylem oleoresin within populations of ponderosa pine," Biochemical Systematics and Ecology, vol. 31, no. 5, pp. 451–465, 2003. · ·

335. E. Sertel, A. Robock, and C. Ormeci, "Impacts of land cover data quality on regional climate simulations," International Journal of Climatology, vol. 30, no. 13, pp. 1942–1953, 2010. · ·

336. G. B. Bonan, S. Levis, L. Kergoat, and K. W. Oleson, "Landscapes as patches of plant functional types: an integrating concept for climate and ecosystem models," Global Biogeochemical Cycles, vol. 16, no. 2, pp. 5–1, 2002.

337. A. Guenther, T. Karl, P. Harley, C. Wiedinmyer, P. I. Palmer, and C. Geron, "Estimates of global terrestrial isoprene emissions using MEGAN (Model of Emissions of Gases and Aerosols from Nature),"Atmospheric Chemistry and Physics, vol. 6, no. 11, pp. 3181–3210, 2006.

338. G. B. Bonan, P. J. Lawrence, K. W. Oleson et al., "Improving canopy processes in the Community Land Model version 4 (CLM4) using global flux fields empirically inferred from FLUXNET data," Journal of Geophysical Research, vol. 116, no. G02, 2011.

339. B. Clement, M. L. Riba, R. Leduc, M. Haziza, and L. Torres, "Concentration of monoterpenes in a maple forest in Quebec," Atmospheric Environment, vol. 24, no. 9, pp. 2513–2516, 1990.

340. C. Geron, A. Guenther, J. Greenberg, T. Karl, and R. Rasmussen, "Biogenic volatile organic compound emissions from desert vegetation of the southwestern US," Atmospheric Environment, vol. 40, no. 9, pp. 1645–1660, 2006.

341. K. Jardine, L. Abrell, S. A. Kurc, T. Huxman, J. Ortega, and A. Guenther, "Volatile organic compound emissions from Larrea tridentata (creosotebush)," Atmospheric Chemistry and Physics, vol. 10, no. 24, pp. 12191–12206, 2010.

342. D. A. Exton, D. J. Suggett, M. Steinke, and T. J. McGenity, "Spatial and temporal variability of biogenic isoprene emissions from

a temperate estuary," Global Biogeochemical Cycles, vol. 26, 2012. ·

343. E. Ormeño, D. R. Gentner, S. Fares, J. Karlik, J. H. Park, and A. H. Goldstein, "Sesquiterpenoid emissions from agricultural crops: correlations to monoterpenoid emissions and leaf terpene content,"Environmental Science and Technology, vol. 44, no. 10, pp. 3758–3764, 2010.

3

Polysaccharides, Proteins, and Phytoplankton Fragments: Four Chemically Distinct Types of Marine Primary Organic Aerosol Classified by Single Particle Spectromicroscopy

Lelia N. Hawkins and Lynn M. Russell

Scripps Institution of Oceanography, University of California, San Diego, La Jolla, CA 92117, USA

ABSTRACT

Carbon-containing aerosol particles collected in the Arctic and southeastern Pacific marine boundary layers show distinct chemical signatures of proteins, calcareous phytoplankton, and two types of polysaccharides in Near-Edge Absorption X-ray Fine Structure (NEXAFS) spectromicroscopy. Arctic samples contained mostly supermicron sea salt cuboids with a polysaccharide-like organic coating. Southeastern Pacific samples contained both continental and marine aerosol types; of the 28 analyzed marine particles, 19 were characterized by sharp alkane and inorganic carbonate peaks in NEXAFS spectra and are identified as fragments of calcareous phytoplankton. Submicron spherical particles with spectral similarities to carbohydrate-like marine sediments were also observed in Pacific samples. In both regions, supermicron amide and alkane-containing particles resembling marine proteinaceous material were observed. These four chemical types provide a framework that incorporates several independent reports of previous marine aerosol observations, showing the diversity of the composition and morphology of ocean-derived primary particles.

INTRODUCTION

The transfer of organic components from the ocean surface to marine aerosol through bubble bursting was shown over 40 years ago [1–3]. These components, referred to as "marine primary organic aerosol" or marine POA [4], have been observed to contribute to organic mass in remote and coastal marine locations [3, 5–8]. In some cases, primary components have been observed to compose greater than 70% of measured submicron OC [6, 7]. The production of submicron particles from bubble bursting remains a key aspect of the global radiation budget because large particle sources are limited to continental and coastal regions [9]; yet the remote marine atmosphere covers more than half of the earth's surface. In remote regions, marine-derived particles have been estimated to account for up to 90% of cloud condensation nuclei (CCN) [10]. Decreases projected for Arctic sea ice extent in response to climate warming may contribute an additional 40–200 ng m^{-3} of aerosol organic carbon (OC) by 2100 from a combination of increased surface ocean productivity and increased spatial extent of wave action [11]. This change in OC is significant considering that background concentrations of less than 1 μg m^{-3} are common in the remote MBL [8, 10, 12–15].

In ocean surface waters, rising bubbles scavenge organic material that is transferred to the atmosphere as the bubble bursts [1, 16, 17]. Much of this scavenged organic material has been classified as "exopolymers," which are mostly composed of polysaccharides [18]. The potential for breaking waves to contribute organic mass to aerosol particles increases with the high concentration of surface active organic compounds and microorganisms enriched in the surface microlayer (SML), relative to the underlying water [19–21]. Observed enrichment factors (EFs) are several orders of magnitude for dissolved and particulate organic carbon (OC) and for specific components like bacteria and viruses. The production of sea spray from bubble bursting results in further enrichment of OC [8, 21, 22]. EFs for organic components in marine aerosol particles have been reported from 5 (viruses and bacteria) to over 100 (organic carbon) from the SML [8, 21]. Since the surface ocean is the primary source of marine POA, the types and relative contributions of organic compounds are expected to be similar. Chemical characterization of the SML and surface water has revealed that carbohydrates constitute 80% of TOC [23], although lipid and protein components have also been observed [21, 24]. Investigations of the composition of airborne marine organic particles have shown multiple lines of evidence for carbohydrates [5, 7, 8, 22, 25–27], amino acids [22, 28], and marine microorganisms [21, 25], confirming that many of the organic components found in the SML and surface ocean are transferred to the marine atmosphere. It is crucial for understanding the role of sea spray aerosol in marine aerosol-cloud interactions that we not only quantify the organic fraction but also characterize its composition, since the hygroscopicity of organic components varies so widely. One important question that remains is how these marine organic components are mixed in airborne particles, since the CCN activity of organic particles can be significantly altered by small amounts of soluble material [29].

To better characterize marine POA in the remote marine boundary layer, aerosol particles were collected during research cruises in the Arctic and southeastern Pacific oceans in local springtime. Single particle X-ray spectromicroscopy was used to separate individual particles into four distinct types of marine POA using organic functional groups, particle morphology, and elemental composition. The findings of this analysis are compared in the context of previous marine POA observations using a variety of analytical techniques.

METHODS

Sample Collection

Ambient aerosol particles for Scanning Transmission X-ray Microscopy with Near-Edge X-ray Absorption Fine Structure (STXM-NEXAFS) analysis were collected in 2008 as part of the International Chemistry Experiment in the Arctic LOwer Troposphere (ICEALOT) and VAMOS Ocean Cloud Atmosphere Land Study Regional Experiment (VOCALS-REx) research cruises, using nearly identical sample collection techniques. The ICEALOT cruise through the North Atlantic and Arctic Oceans was conducted in March and April 2008 on the UNOLS R/V Knorr to investigate the composition and sources of atmospheric aerosol and gas phase species to the northern polar region. Detailed descriptions of the ICEALOT cruise track, sampled air mass histories, and related aerosol measurements are described in [8] and the associated supplementary material. All ICEALOT single particles presented here were collected north of 63°N; most particles were collected within the Arctic Circle (north of 66.56°N). In October and November 2008, the NOAA R/V Ronald Brown traveled in the southeastern Pacific Ocean in the region along 20°S as part of VOCALS-REx, a multiplatform campaign designed to investigate ocean-atmosphere interface processes and to probe aerosol-cloud interactions in the stratocumulus-topped MBL [30]. Details of the VOCALS-REx cruise track, sampled air mass histories, and aerosol chemistry are described in [15]. VOCALS-REx single particle samples were collected along the 20°S portion of the cruise track, including both coastal and remote marine locations. For simplicity, all ICEALOT particles will be referred to as "Arctic» and all VOCALS-REx particles will be referred to as "Pacific.»

Particles were collected through a shared, isokinetic sampling inlet 18 m above sea level [31] and impacted onto silicon nitride windows (Si_3N_4, Silson, Ltd., Northampton, England) at 1 LPM (providing a 2.5 μm 50% efficiency size cut) using a rotating impactor (Streaker, PIXE International Corp., Tallahassee, FL). This impactor was located in a humidity-controlled enclosure; the relative humidity was below 30% during ICEALOT and was controlled at 55% during VOCALS-REx. Windows were sealed and stored frozen until analysis.

Analysis

STXM-NEXAFS

Particles were analyzed on Beamline 5.3.2 at the Advanced Light Source in Lawrence Berkeley National Laboratory (Berkeley, CA) at atmospheric temperature and under dry He (1 atm). Details of STXM-NEXAFS analysis of atmospheric aerosol particles are described in [32, 33], and a brief description is provided here. Image scans from 278 to 320 eV (with up to 0.2 eV resolution) of individual particles provide X-ray absorption spectra of the carbon K-edge, with characteristic peaks from various energy transitions of the bound carbon atoms. Organic and inorganic carbon-containing functional groups are identified by their specific absorption energy between 280 and 320 eV (Table 1). Potassium L-edge transitions also occur in this region. Only particles with measurable difference in absorbance between 280 and 292 eV (the carbon edge) are selected for image scans. Energy calibrations were performed within 48 hours of particle analysis using CO_2 as the reference material. All necessary adjustments were less than 0.05 eV. Absorption spectra from each pixel within the two-dimensional particle image are averaged and normalized following the procedure described in [33]. Spectra normalization entailed subtracting background absorbance (278–283 eV) followed by normalizing to total carbon content (301–305 eV). This normalization provides more uniform spectra for qualitative comparison. Image alignments were performed in Matlab (Mathworks Inc.) using a normalized cross-correlation algorithm implemented in the Matlab image-processing toolbox [33]. An automated algorithm for peak fitting [33] provides relative absorption of aromatic/alkene R(C=C)R', ketone R(C=O)R', alkyl R(C–H)$_n$R', carboxylic carbonyl R(C=O)OH, alcohol R–COH, and carbonate CO_3^{2-} carbon. Spherical-equivalent geometric diameter is used to approximate particle size and is equal to the diameter of a sphere having the same area as the sum of individual pixels with signal above the background level.

Table 1: X-ray spectra carbon K-edge, near-edge, and postedge features

Component	Transition	Energy (eV)
Aromatic/alkene, R(C=C)R'	C 1s-$\pi^*_{C=C}$	284.4–286.4a
Ketone, R(C=O)R'	C 1s-$\pi^*_{C=O}$	286.2–290.9a
Alkyl, R(C–H)nR'	C 1s-σ^*_{CH}	287.4–288.5a
Amide carbonyl, R–NH(C=O)R'	C 1s-$\pi^*_{C=O}$	288.3 ± 0.2b
Carboxylic carbonyl, R(C=O)OH	C 1s-$\pi^*_{C=O}$	288.2–288.9 ± 0.3a
Alcohol, R–COH	C 1s-3p/σ^*_{COH}	289.5 ± 0.3b
Inorganic carbonate, CO32-	C 1s-$\pi^*_{C=O}$	290.4a
Alkyl, R(C–H)nR'	C 1s-σ^*_{CC}	290.8–293a
Potassium, K	L2,3 edges	297.4 ± 0.2 and 299 ± 0.2c

a[34], b[35], c[36].

Individual particle spectra were clustered using a guided Ward clustering algorithm based on a training set of spectra from the 14 particle classes described in [32]. Following clustering, visual inspection of the resulting classes identified 4 spectra types whose class assignments did not accurately represent their spectral features. These spectra had not been observed in previous STXM-NEXAFS studies of atmospheric particles and therefore were not represented in the 14-class training set. The interpretation of these spectra is described in detail in Section 3.

SEM-EDX

Following STXM-NEXAFS analysis, a subset of analyzed carbon-containing single particles (11 particles) were investigated for elemental composition using Scanning Electron Microscopy with Energy Dispersive X-rays (SEM-EDX) at the Scripps Institution of Oceanography Analytical Facility (La Jolla, CA) using a model FEI

Quanta 600 microscope at 10 keV. Samples were uncoated and were analyzed under moderate vacuum. All samples showed Si and N absorption due to the sample substrate. Identified elements include C, O, Ca, S, Na, Mg, and Cl.

RESULTS AND DISCUSSION

Figure 1 shows the distribution of analyzed carbonaceous particles in Pacific and Arctic samples categorized by particle-average spectra. Nonmarine particle types include soil dust, combustion, and secondary particles. These particle types have been observed in previous measurements in urban locations (e.g., Mexico City) and areas affected by urban outflow (e.g., offshore China, the Caribbean, and the Pacific Northwest) [32]. Soil dust particles are characterized by carbonate, potassium, and carboxylic acid-containing organic components (Type "f" in [32]) and are attributed to air masses passing near Santiago and other urban areas along the arid Chilean coast before reaching the ship [15].

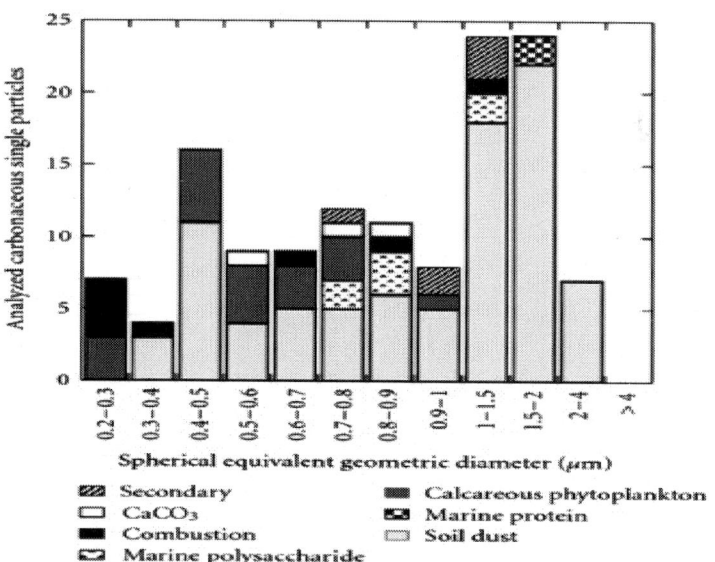

(a)

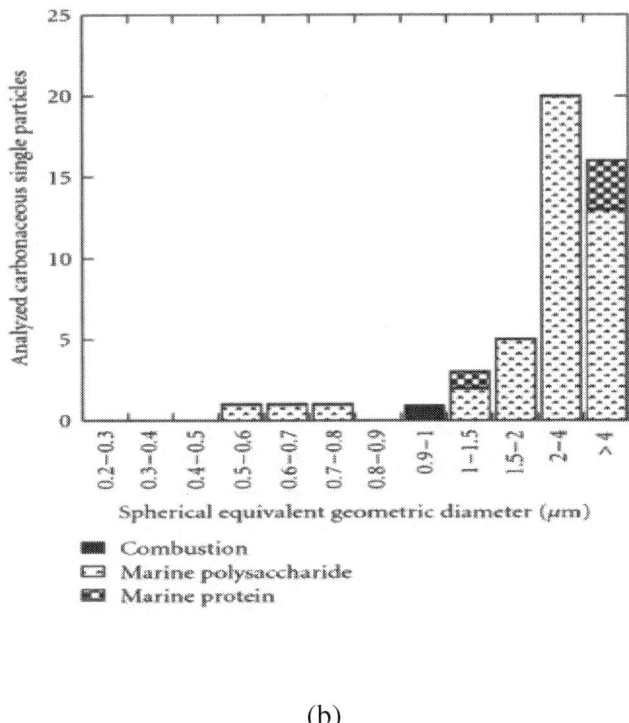

(b)

Figure 1: Distribution of the number of analyzed particles from (a) southeastern Pacific and (b) Arctic marine boundary layers. Particles labeled as "marine polysaccharide" in Arctic samples correspond to Figure 2(a) (PsI) while those in southeastern Pacific samples correspond to Figure 2(b) (PsII). Particle frequency for the collected samples is partly a result of sampling bias and does not represent the observed particle size distribution.

carbon absorption at 292 eV (similar to Type "d" in [32]). With one exception, these particles were submicron, and four out of eight particles were below 300 nm spherical equivalent diameter. Secondary type particles (type "a" in [32]) are characterized by broad carbon absorption beyond 300 eV and by carboxylic carbonyl absorption at 288.7 eV. In previous studies in marine locations, these particles have been the most commonly observed type [32]. In Pacific samples, however, much of the carboxylic acid-containing organic mass is associated with soil dust particles, consistent with measurements reported in [37] of internal mixtures of oxalic and malonic acids with mineral dust.

All organic particles not included in the soil dust, combustion, or secondary particle types were identified as marine origin and fell into four types: carboxylic acid-containing polysaccharides (Arctic), low-solubility polysaccharides (Pacific), calcareous phytoplankton fragments (Pacific), and proteinaceous material (Arctic and Pacific) (Table 2). Marine particles were observed in both Pacific and Arctic samples; however, most of the particles collected in the Arctic region were supermicron. The features and interpretation of the NEXAFS spectra and STXM morphology of particles in each marine type are discussed in detail in the following sections. In addition, three Pacific particles were identified with carbonate and potassium absorption but without any signatures of organic carbon. Their spectra are very similar to type "E" particles found in ocean sediments in [38], which were identified as marine calcium carbonate. These particles are labeled "$CaCO_3$" in Figure 1 but are not included below since they lack organic components.

Table 2: Summary of observed marine particle types in southeast Pacific and Arctic samples

Type	No. of Marine Particles	
	Pacific	Arctic
Polysaccharide		
with carboxylic acid (PsI)	0	43
without carboxylic acid (PsII)	7	0
Protein	2	4
Phytoplankton	19	0
Total	28	47

Carboxylic Acid-Containing Polysaccharides on Sea Salt

Figure 2(a) shows single particle spectra (and category average) for the most commonly observed marine particle type. Spectra in this category have strong carboxylic carbonyl peaks and weak alcohol, carbonate, and potassium peaks. These particles were seen in Arctic samples and compose 43 of the 48 analyzed Arctic particles. Two particles collected at a coastal site in California, which is frequently influenced by marine air masses, also share these features [39]. This particle type is distinct from Type "a" particles in [32] in the stronger contribution of the carboxylic carbonyl peak and the broad alkyl absorption near 293 eV. Filter measurements of submicron particles from the Arctic show a large contribution from alcohol (C–OH) groups to OM attributed to marine carbohydrate-like compounds [8], consistent with previous chemical characterization of the surface microlayer as 80% carbohydrate [23] and with exopolymer secretions (EPSs) repeatedly identified in submicron marine aerosol [7, 25–27]. Just under 90% of the observed Arctic supermicron particles do not show a significant peak at 289.5 eV (C–OH transition), which is different from most of the reported carbohydrate reference spectra [35]. A fraction of these observed spectra do have a shoulder located near 289.5 eV; yet all spectra are dominated by a large peak near 288.7 eV (carboxylic carbonyl). Relative NEXAFS absorption of carboxylic carbonyl and alcohol groups in acid-group-containing reference polysaccharides shows a similar trend; for example, muramic acid and alginic acid show stronger carboxylic carbonyl (π^* transition) peaks than alcohol (σ^* transition) peaks [35] despite the fact that the molar ratio of carboxylic acid to alcohol groups is 0.33 in muramic acid and 0.5 in alginic acid. These compounds are found in bacterial (muramic) and brown algae (alginic) cell walls as structural polysaccharides. Glucuronic acid is another carboxylic acid containing component of polysaccharides that shows strong carboxylic carbonyl absorption (288.65 eV) [35].

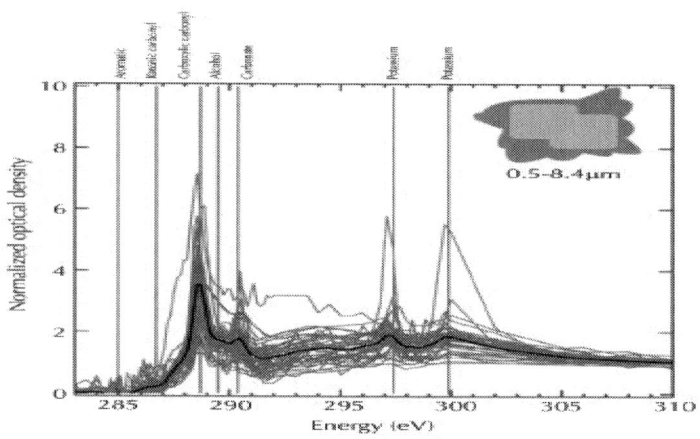

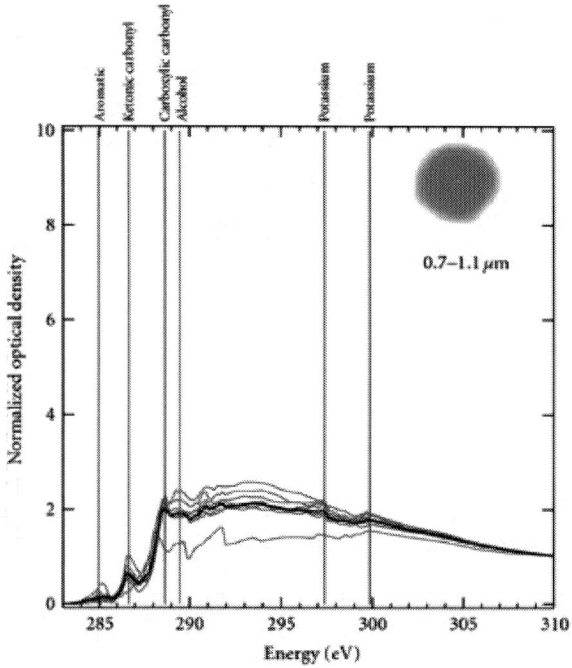

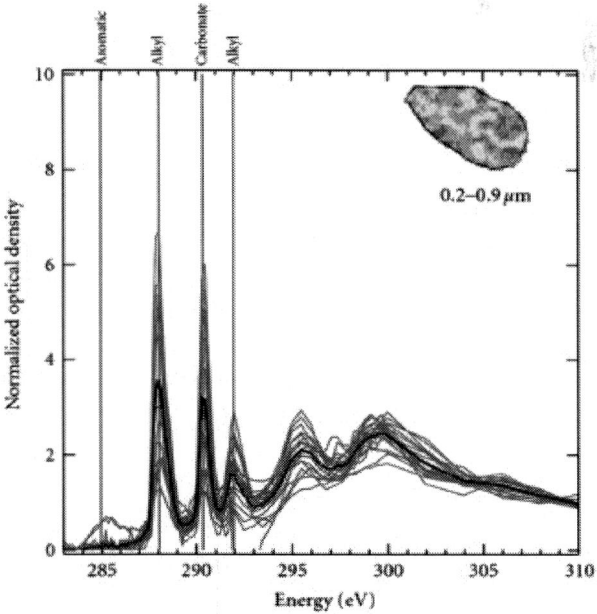

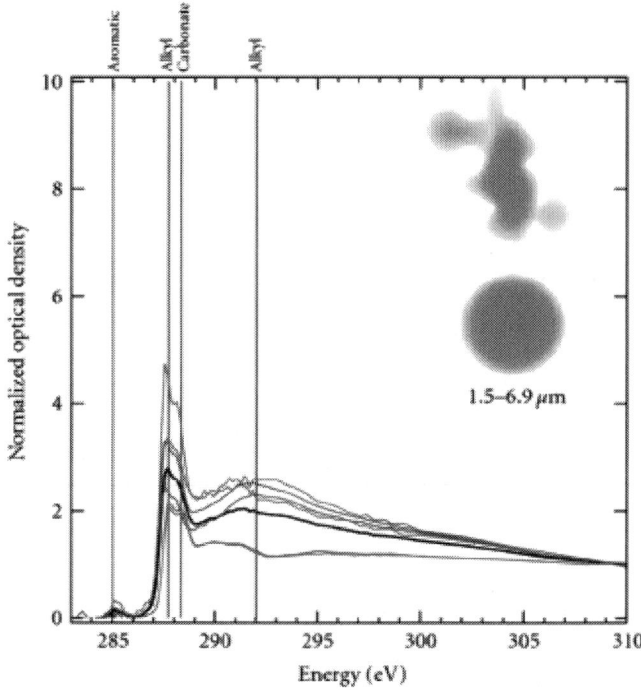

Figure 2: Individual (grey) and average (black) NEXAFS spectra of the four marine particle types including (a) PsI, (b) PsII, (c) calcareous phytoplankton fragments, and (d) proteinaceous particles. Illustrations in each panel represent commonly observed morphologies associated with each spectra type. The observed size range for each type is shown below the illustrations.

Alginic acid is a relevant example for marine POA since the brown algae family includes giant kelp and seaweed found in cold, northern hemisphere oceans [40]. Figure 3 shows the similarities between the average spectrum of particles in this category and an alginic acid reference spectrum [41]. Both spectra show a strong, narrow peak at 288.7 eV and a weaker, broad absorption at 293 eV, without any other organic carbon peaks. Carbonate and potassium absorbances in the average carboxylic acid-type spectrum can be attributed to the sea salt associated with these particles. Many of the particles in this type were characterized by inorganic cuboid structures with an uneven, organic coating (Figure 4(a)), which also were absent in the previously observed secondary particles. EDX spectra of four particles from this class are shown in Figure 5 of Russell et al. [8]; observed peaks include Na and Cl for all four particles. This small amount of organic relative to crystallized sea salt is consistent with the lower organic enrichment expected for supermicron particles rather than submicron particles. This morphology suggests that the organic components on these particles are more soluble than previously reported polysaccharides, which are generally colloidal spherules not associated with sea salt [25–27]. The association with seawater components is also consistent with the assignment of these particles as carboxylic acid-containing polysaccharides like alginic acid, since it has a strong tendency to take up water. These particles are referred to as "Type I polysaccharides" or PsI.

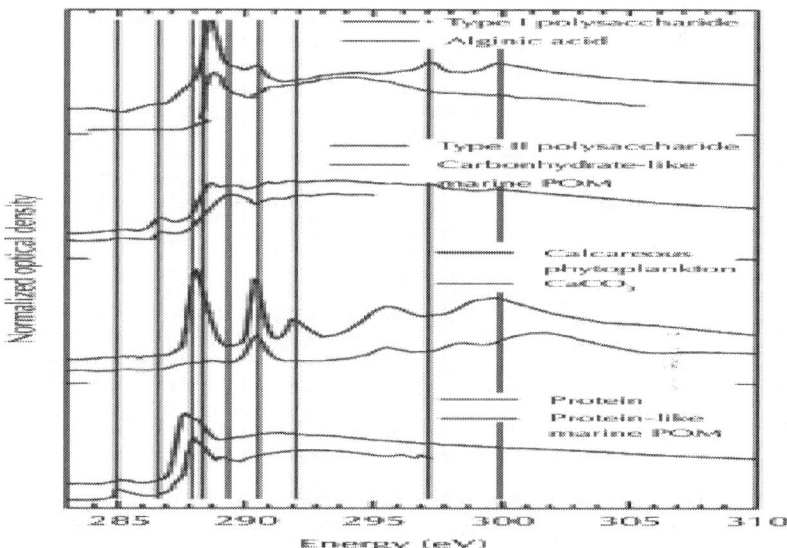

Figure 3: Normalized average spectra for each of the four marine particle types and (below) corresponding reference spectra with similar features. Spectra were reproduced from [41] (alginic acid), [38] (carbohydrate and protein-like marine POM), andhttp://xray1.physics.sunysb.edu/~micros/xas/xas.html, unpublished ($CaCO_3$). Vertical grey lines mark relevant transitions (from left to right): 285 eV ($R(C = C)R'$), 286.7 eV ($R(C=O)R'$), 288.1 eV ($R(C–H)_nR'$), 288.3 eV ($R–NH(C=O)R'$), 289.5 eV ($R–COH$), 290.4 eV (CO_3^{2-}), 292 eV ($R(C–H)_nR'$), 297.4 eV (K), and 299 eV (K).

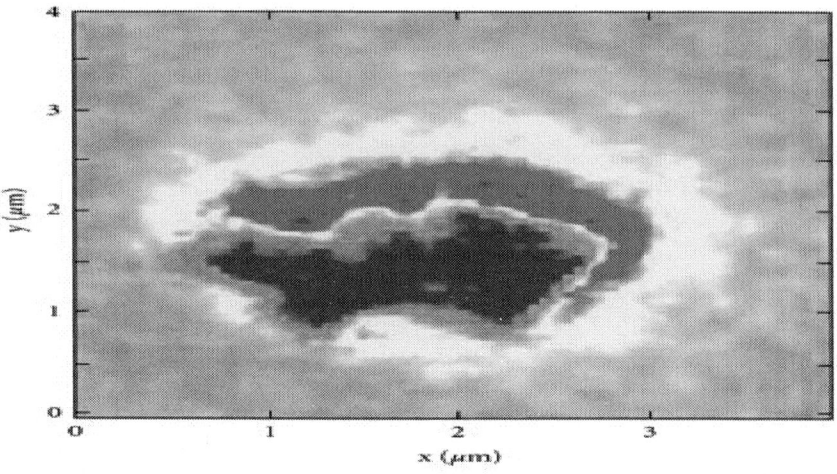

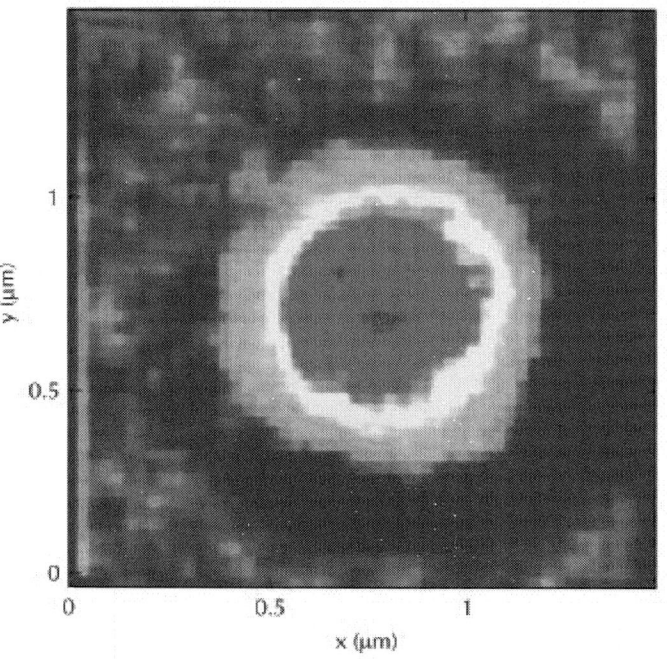

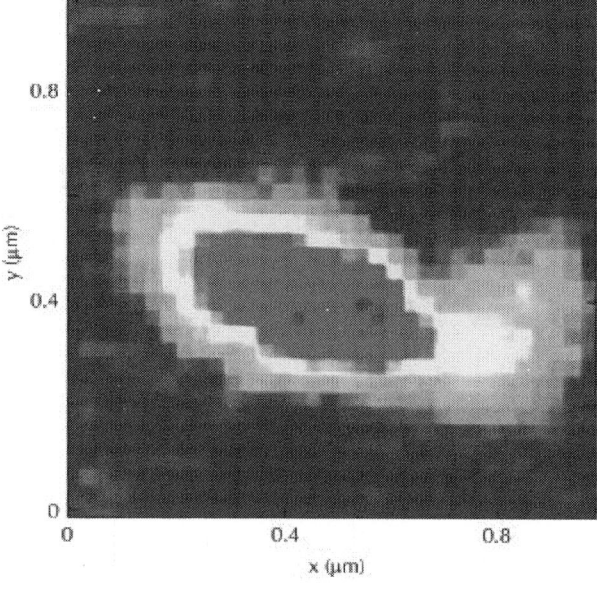

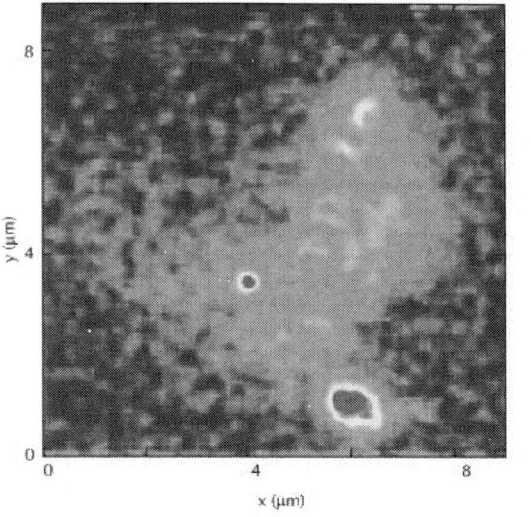

Figure 4: Relative carbon images of representative particles for (a) PsI, (b) PsII, (c) calcareous phytoplankton, and (d) proteinaceous particle types. For each image, the red-blue color scale is relative to individual particle carbon absorption, where red denotes the maximum carbon absorption and blue

denotes the minimum.

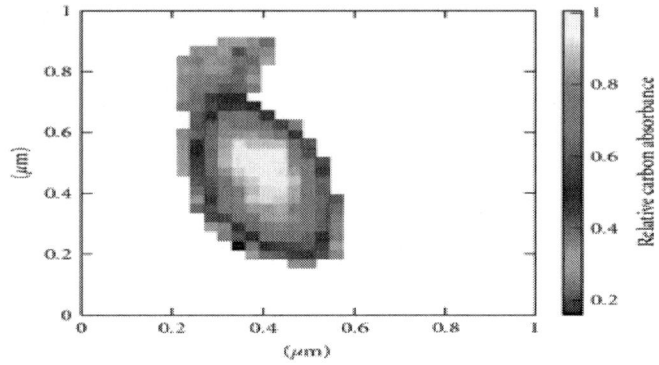

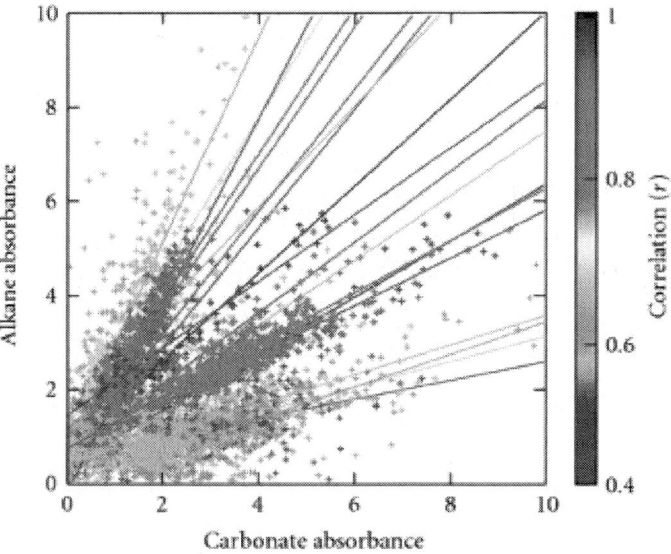

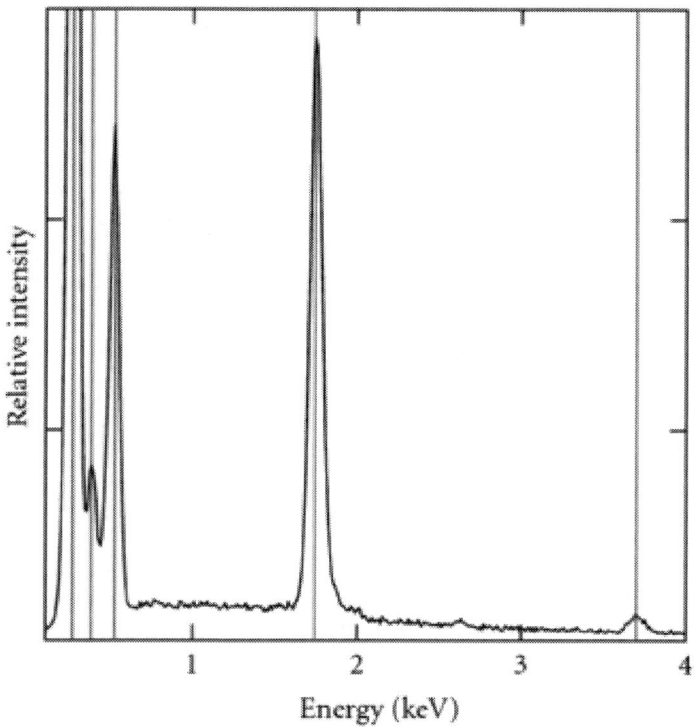

Figure 5: (a) Relative carbon absorbance per pixel from integrated NEXAFS particle-average spectrum for a calcareous phytoplankton fragments. (b) Alkane absorbance compared with carbonate absorbance for pixel-by-pixel fit of NEXAFS spectra of all calcareous phytoplankton type particles. Markers are colored by the correlation coefficient for each least-squares linear regression (one color per particle). (c) EDX spectrum of the same particle at 10 keV accelerating voltage. Vertical red lines mark the C, N, O, Si, and Ca absorbances from left to right. N and Si absorbances are from the sample substrate.

Low-Solubility Polysaccharides

Figure 2(b) shows single particle spectra (and category average) for particles with visible alcohol C–OH absorption (289.5 eV) accompanied by aromatic, ketonic, and carboxylic carbonyl carbon peaks found only in Pacific samples. Here the carboxylic carbonyl absorption is approximately equal to the alcohol carbon absorption.

Reference polysaccharides with equivalent peak heights at or near 288.7 (carboxylic carbonyl) and 289.5 eV (alcohol) include chitin and L-rhamnose [35]. Chitin does not contain any carboxylic carbonyl groups but does contain amide carbonyl groups (monomers are N-acetylglucosamine) which may be responsible for the peak at 288.4 eV. Glucosamine is also present in a 1 : 1 ratio with muramic acid monomers in peptidoglycan, which has been shown to be a major constituent of marine dissolved organic matter (DOM) [43]. Therefore, the observed peak in the average alcohol-type spectrum near 288.7 eV could be attributed either to carbonyl in amide groups or to a mixture with carboxylic carbonyl-containing polysaccharides. It is more probable that these particles contain a mixture of structural polysaccharides than isolated compounds, resulting in less pronounced spectral features than the reference spectra. In fact, the most similar spectrum to the category average comes from a sediment sample of marine particulate organic matter (POM, [38]) (Figure3). [38] used factor analysis to separate different biological compounds in marine POM, and one factor with significant C–OH absorption was identified as carbohydrate material. The carbohydrate-containing marine POM shares the aromatic and ketonic carbon absorbances with the spectra of these particles, while reference (pure) structural polysaccharide spectra in [35] do not. Particles of this type are referred to as "Type II polysaccharides" or PsII.

Filter-based FTIR spectroscopic measurements of Pacific submicron particles show a significant contribution from marine OM (from factor analysis) that is most prominent in sampled air masses with low PM_1 particle mass ($<1 \mu g\, m^{-3}$) and with low radon concentration ($<200\, mBq\, m^{-3}$), indicating little continental influence [15]. Complementary ion chromatography (IC) measurements show low concentrations of submicron Na^+ ($<0.1\ \mu g\, m^{-3}$) or Cl^- ($<0.07\ \mu g\, m^{-3}$), which is consistent with the relatively calm seas encountered during the cruise. PsII particles are spherical, with no cuboidal inorganic core (Figure 4(b)), similar to the spherical colloidal structures observed in TEM by [25–27]. The lack of cuboids is consistent with the lower fraction of Na/OM expected in submicron particles [44].

Calcareous Phytoplankton Fragments

Figure 2(c) shows single particle spectra (and category average) for particles with three strong, narrow peaks at 288.1, 290.4, and 292 eV associated with alkyl R(C–H)nR' (π*), inorganic carbonate CO_3^{2-} (π*), and alkyl R(C–H)nR' (σ*) transitions, respectively. These particles were strictly submicron and found in Pacific samples. A particle with this same characteristic signal was also found in a sample collected at a California coastal site [39]. Compared with all other particle-average spectra, these spectra have much stronger signal-to-noise and have little particle-to-particle variability. These particles also have very little pre-edge absorbance indicating that they are entirely composed of the absorbing (carbonaceous) material, consistent with their strong signal. The narrow alkyl peaks indicate little variation in the neighbors of the absorbing alkyl carbon atoms (e.g., straight-chain alkane compounds) as does the absence of other organic carbon peaks.

The carbonate peak at 290.4 eV is also strong and narrow, indicating that other than the long-chain hydrocarbon compounds, the particle is mostly some form of carbonate. The reference spectrum for $CaCO_3$ is shown in Figure 3. $CaCO_3$ shares the sharp peak at 290.4 eV and the multiple, broad peaks to the right of 295 eV with the average spectrum. To determine the type of carbonate-based mineral, 6 of the 19 particles in this category were analyzed with SEM-EDX; all particles showed strong C, O, and Ca signals while S, Na, Mg, and Cl were absent or weak (Figure 5(c)). These particles show a variety of nonspherical shapes. Some particles appear elliptical with sharp points (Figure 4(c)) and others are amorphous. Based on their appearance, the particles resemble small, dust-like fragments. However, their chemical composition is not consistent with aged or processed dust transported to the remote MBL. In addition, long-chain hydrocarbons are not typical of secondary organic aerosol [45]; the absence of S in EDX spectra also makes it unlikely that atmospheric processing is responsible for the majority of organic mass in these particles.

Previous observations of excess Ca^{2+}, relative to sea salt ratios, in marine aerosol have been attributed to fragments of calcium carbonate-producing phytoplankton (coccolithophores) emitted to the atmosphere during bubble bursting [46]. Other possible sources of elevated calcium include EPS, which have been shown to incorporate calcium in gel formation [47]. These single-celled phytoplankton produce delicate, calcium carbonate scales (coccoliths) that continually slough off the organisms during their growth and that are released during predation [48]. These scales are oval-shaped and are typically 500–3000 nm in length, resulting in fragments that are consistent with the observed size range of these alkane/carbonate particles. Coccolithophores (especially Emiliania huxleyi) are abundant in both high- and low-latitude oceans and are responsible for about half of the total oceanic carbonate production [49]. Their blooms are so large and persistent that they can been seen from space in satellite images of ocean color as patches of light green against the dark blue ocean. A recent study measuring whole coccolithophores, detached scales, and calcite fragments in surface waters in the same region as the VOCALS-REx cruise has documented their abundance in the Peru-Chile Upwelling (PCU) and the South Pacific Gyre (SPG) [49]. The measured seawater carbonate particle surface area distribution in their work showed a large peak between 2 and μ3 m (corresponding to whole coccoliths with diameters between 1.6 and 2μ m) and a smaller peak at 250 nm (corresponding to coccolith fragments with diameters around 560 nm). This smaller mode is consistent with the size range of observed particles in this category.

In addition to producing a large fraction of oceanic carbonate, coccolithophores are known to produce extremely stable, lipid-like compounds called alkenones (nC_{37}–C_{39}), which contain one ketone group and two or three degrees of unsaturation [50]. Although the exact function of these compounds is unknown, an investigation of alkenones in various organelles and membranes of Emiliania huxleyi has shown that they are predominantly located in the coccolith-producing compartment (CPC) of the cell and are most likely membrane-unbound lipids associated with the function of the CPC [51]. The coproduction of these long-chain alkanes with calcite coccoliths is consistent with the strong, sharp alkyl peaks present in our alkane/carbonate particle spectra and with the absence of other groups, such as carboxylic acids. Coproduction would also result in a similar ratio of the two species (alkane and carbonate) over the particle, rather than separate carbonate and alkane-dominated regions. Figure 5(b) shows the pixel-by-pixel normalized alkane absorption compared with normalized carbonate absorption for each of the 19 alkane/carbonate-type particles. Correlations between these two groups are strong (12 of the 19 particles have r>0.75). These strong correlations demonstrate the uniformity of the two groups over individual particles, though the relative amounts of alkane and carbonate groups (i.e., the fitted slopes) vary among particles. Given these observations, the alkane/carbonate particles will be referred to as "Calcareous phytoplankton fragments" in the remaining sections.

Proteinaceous Particles

Figure 2(d) shows single particle spectra (and category average) for particles with aromatic/alkene, alkyl, and amide carbon absorptions at 285, 287.7, and 288.2 eV, respectively. The aromatic/alkene peak at 285 eV has a shoulder at 285.4 eV in all 6 particles indicating the presence of multiple unsaturated carbon environments. These spectra, like the calcareous phytoplankton spectra, have low noise and are quite similar to one another in terms of peak locations, shapes, and relative peak heights. Unlike the other categories, particles with this signature are found in both Arctic and Pacific samples but with slightly different morphologies. The two Pacific particles are spherical and all four of the Arctic particles are loose agglomerations of carbonaceous material (Figure 4(d)). The most unique feature of these spectra is the shoulder at 288.2 eV, corresponding to carbonyl carbon in an amide group [34, 35]. Amide groups have also been identified from the CNH σ^* transition at 289.5 eV [34, 52]. Amide groups (known as peptide bonds when found in proteins) are formed from dehydration reactions of the carboxylic acid group of one amino acid monomer and the amine group of another. Therefore, reference spectra for amino acids that have strong carboxyl carbonyl absorption [35] are not representative of bound amino acid monomers in proteins. The broad alkyl absorption near 292 eV indicates that a variety of alkyl carbon environments exist in these particles, contrasting the sharp peak at 292 eV in the calcareous phytoplankton fragments. In addition, the presence of two alkyl carbon peaks and the absence of the carboxylic carbonyl peak indicate that these proteinaceous compounds may be related to lipoproteins that are found in the membranes of chloroplasts. Lipoproteins contain both lipid and protein components and could be responsible for the significant alkyl absorption seen here. Aromatic and alkene groups are found in proteins as well. Phenylalanine, tyrosine, histidine, and tryptophan are all amino acids with aromatic or alkene side groups.

The fourth pair of spectra in Figure 3 show the spectral similarities between the average amide-type particle spectrum and the protein-like component of marine POM identified in [38]. The two spectra share the small shoulder at 285.4 eV and the amide and broad alkyl absorption regions. However, the amide-type average spectrum has more π^* alkyl absorption (287.7 eV) (which is associated with long-chain hydrocarbons such as lipids) than the protein-like marine POM. The lipid component may give these particles more surface active properties and may result in preferential concentration in the surface microlayer. If this is the case, lipid-containing proteinaceous material would be preferentially transferred to the atmosphere during bubble bursting over nonlipid proteinaceous compounds. The particle images in this type, both spherical and agglomerative, show little evidence of sea salt, which is consistent with hydrophobic organic material. In collocated filter measurements of both Pacific and Arctic MBL air masses, primary amines composed 8% of marine OM (from factor analysis). In fact, primary amine groups have been identified in marine OM factors from all ambient measurements where marine factors were identified [8]. That the Pacific and Arctic proteinaceous POA spectra are indistinguishable reflects the apparent chemical similarity of the protein components in marine POA.

Reconciling Marine POA Observations

Over 10 years of measurements of marine POA are summarized in Table 3; although the collection encompasses particle properties determined from diverse techniques from TEM-EDX to HNMR, most observations can be assigned to one of three main types: (1) polysaccharides, (2) proteins and amino acids, or (3) microorganisms and their fragments. Figure 6 illustrates the three main types and their surface ocean counterparts using the four types of marine POA particles observed in this study. The chemical characterization of single marine POA particles suggests that biogenic organic components and microorganisms observed in this and previous studies are present as an external mixture including—but not limited to—polysaccharides, proteins, and microorganisms.

Table 3: Observed types of marine primary organic aerosol and the suggested biological relevance of specific particle types

Location	Method(s)	Particle Size	Dominant Component(s) or Spectral Feature	Biological Relevance
Polysaccharides				
Arctica	TEM,	<100 nm	Colloidal spherules	EPS gels
Variousb	TEM	<1 µm	Colloidal spherules	EPS gels
	X-ray backscatter,			
	and solubility			
Mediterranean Sea and	Alcian blue dye	1–50 µm	Semi-transparent colloids	Polysaccharides
Long Island Soundc				
W. Pacificd	TEM,	<50 nm	Colloidal spherules	EPS gels
	SEM with X-ray backscatter,			
	and solubility			
North Atlantice,*	HNMR (WSOC and WIOC)	60–1000 nm	Hydroxylate aliphatics	Lipopolysaccharides
			Lipid-like aliphatics	
Arcticf	FTIR spectroscopy	<1 µm	Organic hydroxyl groups	Polysaccharides
			Alkane groups	
SE Pacificg	FTIR spectroscopy	<1 µm	Organic hydroxyl groups	Polysaccharides
SE Pacifich	STXM-NEXAFS	<1 µm	Organic hydroxyl groups	Polysaccharides

Arctich	STXM-NEXAFS	>1 µm	Carboxylic acid groups	Polysaccharides
Protein and amino acid compounds				
Arctici	TEM and Extraction	>50 nm	Hydrophobic organic	Amino acids
			aggregates	
Mediterranean Sea and	HPLC and	not provided	Asp, Glu, Ser, Ala	Amino acids
Long Island Soundc	Coomassie Blue dye	1–50 µm	Semi-transparent colloids	Proteins
Arctich	STXM-NEXAFS	>1 µm	Alkane and	Protein
			amide groups	
SE Pacifich	STXM-NEXAFS	>1 µm	Alkane and	Protein
			amide groups	
Micro-organisms and their fragments				
Arctici	TEM and Extraction	400 nm		Bacteria and diatoms
Arctica	TEM,	200–5000 nm		Micro-organisms and
				fragments
	X-ray backscatter,			
	and solubility			
W. Pacificd	TEM,	>400 nm	CaCO3	Coral-related
	SEM with X-ray backscatter,	3.7 to 7.5 µm		Bacteria
	and solubility			
SE Pacifich	STXM-NEXAFS	<1 µm	CaCO3 and alkane groups	Calcareous phytoplankton
				fragments
None listed				
Tasmaniaj	PALMS	>160 nm	Organic mass fragments	
Arctick	TEM	>100 nm	Organic liquid	Proteins
Irelandl	IC,	<1.5µm	WIOC (not characterized)	

	EGA,		WSOC (aliphatic groups	
	HNMR,		near heteroatoms, HULIS,	
	and TOC		and partially oxidized	
			species)	

HNMR characterized aerosol was generated in a laboratory setting from collected seawater.

[a][25], [b][26], [c][22], [d][27], [e][7], [f][8], [g][15], This work, [h][28], [i][14], [j][42], [k][6].

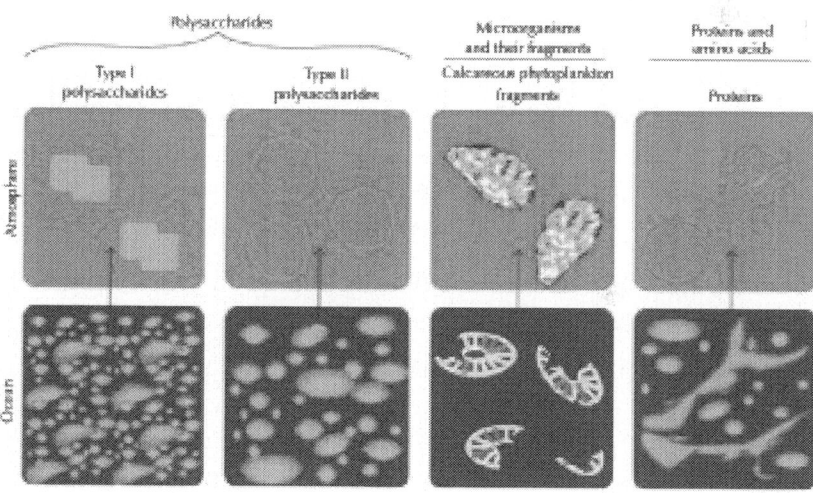

Figure 6: Illustration of the four observed marine particle types in the ocean and atmosphere.

Using TEM images of colloidal spherules, X-ray backscatter of elemental components, and tests for solubility, Leck and Bigg [25, 26] deduced that the hydrated, heat-resistant, hydrophobic organic substance present in submicron marine aerosol was related to exopolymer secretions (EPSs), which are high molecular weight, hydrated polysaccharides. This finding was consistent with numerous reports of large concentrations of EPS in surface ocean water [18, 47]. The colloidal structure of the hydrophobic particles is also consistent with the observation of gel formation from the marine EPS [47, 53]. Although the attributes of their measurements of particle shape, size, and solubility were consistent with EPS characteristics, little chemical evidence was available to confirm their composition as polysaccharides. Near the same time, ambient marine particles from the Mediterranean and Atlantic were shown to contain polysaccharide-rich gels using Alcian blue dye, a stain sensitive to all types of polysaccharides [22]. EI-MS measurements of marine aerosol in the western Pacific also showed substantial contributions from carbohydrates (i.e., levoglucosan and glucose) partially attributed to organics from the ocean surface [5]. A subsequent HNMR study of laboratory-generated aerosol (using North Atlantic seawater) corroborated the presence of polysaccharide-like organic components in marine POA by reporting aliphatic and hydroxylated functional groups in addition to lipid-like signatures [7]. The authors proposed lipopolysaccharides as a possible explanation for the observed groups. Evidence that polysaccharides accounted for 44–61% of marine submicron OM was provided in Russell et al. [8] using FTIR spectroscopy. Their work used the chemical similarity of alcohol C–OH groups in ambient marine submicron aerosol with reference FTIR spectra of 11 different polysaccharides (e.g., pectin, glucose, and xylose). That study was the first to report large quantities of specific signatures of polysaccharides associated with sea salt in submicron ambient marine aerosol, consistent with both the physical attributes reported in [25, 27] and the chemical signatures of simulated marine aerosol in [7]. Using single particle spectromicroscopy, we have observed that polysaccharide-containing particles make up a majority of the measured carbonaceous single particles in two marine regions. From these single particle measurements we have also estimated the mass distribution of Arctic and Pacific marine particles, using the spherical equivalent diameter approximated for each particle and an average density of $1 \, \mu g \, m^{-3}$ (for simplicity). Figure 7 shows the combined, approximate mass distribution of Arctic and Pacific marine POA. Together, Type I and II polysaccharides compose 57% of measured submicron particle mass (and 83% of total particle mass), consistent with the observations of [8]. We also show that multiple types of polysaccharides, including water-insoluble compounds resembling chitin, exist in airborne marine particles.

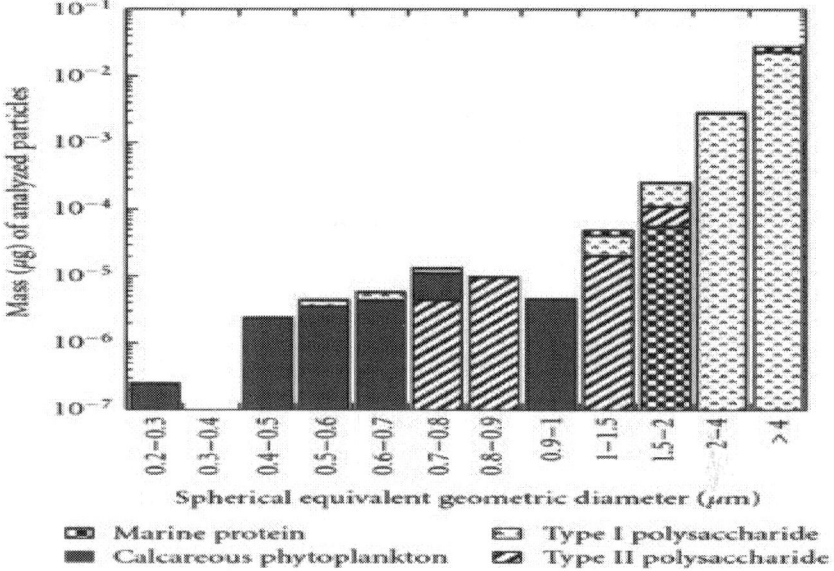

Figure 7: Estimated mass of marine particle types from both Arctic and Pacific samples.

Prior to the discovery of polysaccharides in marine aerosol, TEM analysis of Arctic submicron aerosol particles indicated that the spherical, hydrophobic organic particles could be related to amino acids (i.e., L-methionine) based on the surface active nature of the aerosol particles and on measurements of surface active proteins being scavenged by bubbles in seawater [28]. However, the same properties attributed to proteins in Leck and Bigg [28] could also be attributed to EPS [25–27]. More recently, Kuznetsova et al. [22] used Coomassie blue dye to confirm that some of the colloidal gel-like material surrounding bacteria and virus in Mediterranean and Atlantic marine aerosol samples was indeed proteinaceous. Here we report observations of amide-containing hydrophobic marine aerosol particles from two distant ocean environments that match the characteristic spectral signatures of proteinaceous marine POM, indicating that protein-like organic compounds also contribute to marine POA in many parts of the marine atmosphere.

Marine microorganisms clearly play a large role in marine aerosol formation and composition. In addition to secreting nonvolatile organic components (e.g., polysaccharides, lipids, and proteins) and emitting gas phase precursors to marine aerosol (e.g., dimethyl sulfide, DMS), they can themselves be lofted to the atmosphere where they can serve as surfaces for heterogeneous reactions and as cloud condensation nuclei [1, 21, 25, 28]. Most observations of airborne microorganisms have reported bacteria or diatom fragments, mostly because these particles have distinct shapes easily discernible from other particles in TEM images. Submicron fragments, especially if mixed with gel-like organic material concentrated in the surface microlayer, are extremely difficult to identify based solely on morphology. SEM coupled with EDX can confirm the presence of C, O, and nutrient-affiliated elements like N and P but cannot provide the chemical specificity needed to identify the components of intact cell walls, chloroplasts, or other organelles. For this, X-ray spectromicroscopy is well-suited [38, 41]. Using STXM-NEXAFS we have identified submicron fragments of calcareous phytoplankton (coccolithophores) previously suggested to contribute significant quantities of nss-Ca in MBL aerosol [46]. The unique signature of $CaCO_3$ coupled with straight-chain alkane groups in the average spectra was combined with subparticle resolution spectra—confirming the uniform distribution of the two components—to support the classification of these particles as biological.

CONCLUSIONS

Ambient sub- and supermicron marine aerosol particles were collected in Pacific and Arctic marine boundary layers and subsequently analyzed using single particle STXM-NEXAFS, revealing four distinct types of marine POA. Although two-thirds of marine particles were characterized as polysaccharides, important differences exist even among those seemingly similar biogenic compounds, including the association with sea salt and the inferred differences in hygroscopicity. We also report evidence of proteinaceous compounds and the first observation of calcifying phytoplankton in marine POA.

In previous chemical characterizations of marine aerosol, most observations of marine POA show either hydrophobic, polysaccharide-like material or morphologically distinct microorganisms (i.e., bacteria and diatoms). The particles presented here, while consistent with those observations, provide a more detailed, chemically specific picture of marine aerosol that resolves some of the uncertainties associated with previous observations. These observations also confirm that multiple, distinct types of marine particles are emitted to the atmosphere as external mixtures.

ACKNOWLEDGMENTS

This work was supported by NSF Grant ATM-0744636. The authors thank George Flynn for providing the calcium carbonate reference spectrum. The authors would like to acknowledge Satoshi Takahama and Shang Liu for contributing to the analysis of single particles by STXM-NEXAFS and David Kilcoyne at Beamline 5.3.2 for technical assistance with beamline operation. They would also like to thank Derek Coffman, James Johnson, Drew Hamilton, and Catherine Hoyle for their assistance in sample collection and analysis as well as the captain and crew of the NOAA R/V Ronald Brown and the UNOLS R/V Knorr for their support in the field.

REFERENCES

1.   D. C. Blanchard, "Sea-to-air transport of surface active material," Science, vol. 146, no. 3642, pp. 396–397, 1964.

2.   E. J. Hoffman and R. A. Duce, "Factors influencing the organic carbon content of marine aerosols: a laboratory study," Journal of Geophysical Research, vol. 81, no. 21, pp. 3667–3670, 1976.

3.   R. B. Gagosian, O. C. Zafiriou, E. T. Peltzer, and J. B. Alford, "Lipids in aerosols from the tropical North Pacific: temporal variability," Journal of Geophysical Research, vol. 87, no. 13, pp. 11113–11144, 1982.

4.   M. Kanakidou, J. H. Seinfeld, S. N. Pandis et al., "Organic aerosol and global climate modelling: a review," Atmospheric Chemistry and Physics, vol. 5, no. 4, pp. 1053–1123, 2005.

5.   K. K. Crahan, D. A. Hegg, D. S. Covert et al., "Speciation of organic aerosols in the tropical mid-Pacific and their relationship to light scattering," Journal of the Atmospheric Sciences, vol. 61, no. 21, pp. 2544–2558, 2004.

6.   F. Cavalli, M. C. Facchini, S. Decesari et al., "Advances in characterization of size-resolved organic matter in marine aerosol over the North Atlantic," Journal of Geophysical Research D, vol. 109, no. 24, pp. 1–14, 2004.

7.   M. C. Facchini, M. Rinaldi, S. Decesari et al., "Primary submicron marine aerosol dominated by insoluble organic colloids and aggregates," Geophysical Research Letters, vol. 35, no. 17, p. 5, 2008.

8.   L. M. Russell, L. N. Hawkins, A. A. Frossard, P. K. Quinn, and T. S. Bates, "Carbohydrate-like composition of submicron atmospheric particles and their production from ocean bubble bursting,"Proceedings of the National Academy of Sciences of the United States of America, vol. 107, no. 15, pp. 6652–6657, 2010.

9.   C. H. Twohy, M. D. Petters, J. R. Snider et al., "Evaluation of the aerosol indirect effect in marine stratocumulus clouds: droplet number, size, liquid water path, and radiative impact," Journal of Geophysical Research D, vol. 110, no. 8, pp. 1–16, 2005.

10.   A. D. Clarke, S. R. Owens, and J. Zhou, "An ultrafine sea-salt flux from breaking waves: Implications for cloud condensation nuclei in the remote marine atmosphere," Journal of Geophysical Research D, vol. 111, no. 6, pp. 1–2, 2006.

11.   A. Ito and M. Kawamiya, "Potential impact of ocean ecosystem changes due to global warming on marine organic carbon aerosols," Global Biogeochemical Cycles, vol. 24, p. 10, 2010.

12.   D. M. Murphy, J. R. Anderson, P. K. Quinn et al., "Influence of sea-salt on aerosol radiative properties in the Southern Ocean marine boundary layer," Nature, vol. 392, no. 6671, pp. 62–65, 1998.

13.   P. K. Quinn, D. J. Coffman, V. N. Kapustin, T. S. Bates, and D. S. Covert, "Aerosol optical properties in the marine boundary layer during the First Aerosol Characterization Experiment (ACE 1) and the underlying chemical and physical aerosol properties," Journal of Geophysical Research D, vol. 103, no. 13, pp. 16547–16563, 1998.

14.   A. M. Middlebrook, D. M. Murphy, and D. S. Thomson, "Observations of organic material in individual marine particles at Cape Grim during the First Aerosol Characterization Experiment

(ACE 1)," Journal of Geophysical Research D, vol. 103, no. 13, pp. 16475–16483, 1998.

15. L. N. Hawkins, L. M. Russell, D. S. Covert, P. K. Quinn, and T. S. Bates, "Carboxylic acids, sulfates, and organosulfates in processed continental organic aerosol over the southeast Pacific Ocean during VOCALS-REx 2008," Journal of Geophysical Research, vol. 115, p. 16, 2010.

16. D. C. Blanchard, "Bubble scavenging and the water-to-air transfer of organic material in the sea,"Advances in Chemistry Series, vol. 145, pp. 360–387, 1975.

17. R. Tseng, J. T. Viechnicki, R. A. Skop, and J. W. Brown, "Sea-to-air transfer of surface-active organic compounds by bursting bubbles," Journal of Geophysical Research, vol. 97, no. 4, pp. 5201–5206, 1992.

18. A. W. Decho, "Microbial exopolymer secretions in ocean environments—their role (s) in food webs and marine processes," Oceanography and Marine Biology, vol. 28, pp. 73–153, 1990.

19. S. M. Henrichs and P. M. Williams, "Dissolved and particulate amino acids and carbohydrates in the sea surface microlayer," Marine Chemistry, vol. 17, no. 2, pp. 141–163, 1985.

20. M. Kuznetsova and C. Lee, "Dissolved free and combined amino acids in nearshore seawater, sea surface microlayers and foams: influence of extracellular hydrolysis," Aquatic Sciences, vol. 64, no. 3, pp. 252–268, 2002.

21. J. Y. Aller, M. R. Kuznetsova, C. J. Jahns, and P. F. Kemp, "The sea surface microlayer as a source of viral and bacterial enrichment in marine aerosols," Journal of Aerosol Science, vol. 36, no. 5-6, pp. 801–812, 2005.

22. M. Kuznetsova, C. Lee, and J. Aller, "Characterization of the proteinaceous matter in marine aerosols,"Marine Chemistry, vol. 96, no. 3-4, pp. 359–377, 2005. · ·

23. L. I. Aluwihare, D. J. Repeta, and R. F. Chen, "A major biopolymeric component to dissolved organic carbon in surface sea water," Nature, vol. 387, no. 6629, pp. 166–169, 1997.

24. K. Larsson, G. Odham, and A. Södergren, "On lipid surface films on the sea. I. A simple method for sampling and studies of composition," Marine Chemistry, vol. 2, no. 1, pp. 49–57, 1974.

25. C. Leck and E. K. Bigg, "Biogenic particles in the surface microlayer and overlaying atmosphere in the central Arctic Ocean during summer," Tellus B, vol. 57, no. 4, pp. 305–316, 2005.

26. C. Leck and E. K. Bigg, "Source and evolution of the marine aerosol—a new perspective," Geophysical Research Letters, vol. 32, no. 19, pp. 1–4, 2005.

27. C. Leck and E. K. Bigg, "Comparison of sources and nature of the tropical aerosol with the summer high Arctic aerosol," Tellus. Series B, vol. 60, no. 1, pp. 118–126, 2008.

28. C. Leek and E. Keith Bigg, "Aerosol production over remote marine areas—a new route," Geophysical Research Letters, vol. 26, no. 23, pp. 3577–3580, 1999.

29. G. C. Roberts, P. Artaxo, J. Zhou, E. Swietlicki, and M. O. Andreae, "Sensitivity of CCN spectra on chemical and physical properties of aerosol: a case study from the Amazon Basin," Journal of Geophysical Research D, vol. 107, no. 20, pp. 1–37, 2002.

30. R. Wood, C. Bretherton, B. Huebert, C. R. Mechoso, and R. Weller, "VOCALS SouthEast Pacific Regional Experiment (REx)," Scientific program overview, 2006, http://www.usclivar.org/science_status/VOCALS_SPO_Revised_Complete.pdf.

31. T. S. Bates, P. K. Quinn, and D. Coffman, "Boundary layer aerosol chemistry during TexAQS/GoMACCS 2006: insights into aerosol sources and transformation processes," Journal of Geophysical Research, vol. 113, p. 18, 2008.

32. S. Takahama, S. Gilardoni, L. M. Russell, and A. L. D. Kilcoyne, "Classification of multiple types of organic carbon composition in atmospheric particles by scanning transmission X-ray microscopy analysis," Atmospheric Environment, vol. 41, no. 40, pp. 9435–9451, 2007.

33. S. Takahama, S. Liu, and L. M. Russell, "Coatings and clusters of carboxylic acids in carbon-containing atmospheric particles from spectromicroscopy and their implications for cloud-nucleating and optical properties," Journal of Geophysical Research, vol. 115, 2010.

34. S. C. B. Myneni, "Soft X-ray spectroscopy and spectromicroscopy studies of organic molecules in the environment," Reviews in Mineralogy and Geochemistry, vol. 49, no. 1, pp. 485–579, 2002.

35.   D. Solomon, J. Lehmann, J. Kinyangi et al., "Carbon (1s) NEXAFS spectroscopy of biogeochemically relevant reference organic compounds," Soil Science Society of America Journal, vol. 73, no. 6, pp. 1817–1830, 2009.

36.   T. H. Yoon, K. Benzerara, S. Ahn, R. G. Luthy, T. Tyliszczak, and G. E. Brown Jr., "Nanometer-scale chemical heterogeneities of black carbon materials and their impacts on PCB sorption properties: soft X-ray spectromicroscopy study," Environmental Science and Technology, vol. 40, no. 19, pp. 5923–5929, 2006.

37.   R. C. Sullivan and K. A. Prather, "Investigations of the diurnal cycle and mixing state of oxalic acid in individual particles in Asian aerosol outflow," Environmental Science and Technology, vol. 41, no. 23, pp. 8062–8069, 2007.

38.   J. A. Brandes, C. Lee, S. Wakeham et al., "Examining marine particulate organic matter at sub-micron scales using scanning transmission X-ray microscopy and carbon X-ray absorption near edge structure spectroscopy," Marine Chemistry, vol. 92, no. 1–4, pp. 107–121, 2004.

39.   S. Liu, D. A. Day, and L. M. Russell, "Afternoon increase of oxygenated organic functional groups at a coastal site in Southern California," in preparation.

40.   C. van den Hoek, "Phytogeographic distribution groups of benthic marine algae in the North Atlantic Ocean. A review of experimental evidence from life history studies," Helgoland Marine Research, vol. 35, no. 2, pp. 153–214, 1982.

41.   J. R. Lawrence, G. D. W. Swerhone, G. G. Leppard et al., "Scanning transmission X-ray, laser scanning, and transmission electron microscopy mapping of the exopolymeric matrix of microbial biofilms,"Applied and Environmental Microbiology, vol. 69, no. 9, pp. 5543–5554, 2003.

42.   E. K. Bigg and C. Leck, "Properties of the aerosol over the central Arctic Ocean," Journal of Geophysical Research D, vol. 106, no. 23, pp. 32101–32109, 2001.

43.   R. Benner and K. Kaiser, "Abundance of amino sugars and peptidoglycan in marine particulate and dissolved organic matter," Limnology and Oceanography, vol. 48, no. 1, pp. 118–128, 2003.

44. C. Oppo, S. Bellandi, N. Degli Innocenti et al., "Surfactant components of marine organic matter as agents for biogeochemical fractionation and pollutant transport via marine aerosols," Marine Chemistry, vol. 63, no. 3-4, pp. 235–253, 1999.

45. Q. Zhang, J. L. Jimenez, M. R. Canagaratna et al., "Ubiquity and dominance of oxygenated species in organic aerosols in anthropogenically-influenced Northern Hemisphere midlatitudes," Geophysical Research Letters, vol. 34, no. 13, p. 6, 2007.

46. H. Sievering, J. Cainey, M. Harvey, J. McGregor, S. Nichol, and P. Quinn, "Aerosol non-sea-salt sulfate in the remote marine boundary layer under clear-sky and normal cloudiness conditions: ocean-derived biogenic alkalinity enhances sea-salt sulfate production by ozone oxidation," Journal of Geophysical Research D, vol. 109, no. 19, pp. 1–12, 2004.

47. P. Verdugo, A. L. Alldredge, F. Azam, D. L. Kirchman, U. Passow, and P. H. Santschi, "The oceanic gel phase: a bridge in the DOM-POM continuum," Marine Chemistry, vol. 92, no. 1–4, pp. 67–85, 2004.

48. R. H. M. Godoi, K. Aerts, J. Harlay et al., "Organic surface coating on Coccolithophores—Emiliania huxleyi: its determination and implication in the marine carbon cycle," Microchemical Journal, vol. 91, no. 2, pp. 266–271, 2009.

49. L. Beaufort, M. Couapel, N. Buchet, H. Claustre, and C. Goyet, "Calcite production by coccolithophores in the south east Pacific Ocean," Biogeosciences, vol. 5, no. 4, pp. 1101–1117, 2008.

50. R. W. Jordan and A. Kleijne, "A classification system for living coccolithophores," in Coccolithophores, pp. 83–105, Cambridge University Press, Cambridge, UK, 1994.

51. K. Sawada and Y. Shiraiwa, "Alkenone and alkenoic acid compositions of the membrane fractions of Emiliania huxleyi," Phytochemistry, vol. 65, no. 9, pp. 1299–1307, 2004.

52. L. M. Russell, S. F. Maria, and S. C. B. Myneni, "Mapping organic coatings on atmospheric particles," Geophysical Research Letters, vol. 29, no. 16, p. 4, 2002.

53. W.-C. Chin, M. V. Orellana, and P. Verdugo, "Spontaneous assembly of marine dissolved organic matter into polymer gels," Nature, vol. 391, no. 6667, pp. 568–572, 1998.

Chapter 4

Grand Challenges in Atmospheric Science

Luis Gimeno

EPhysLab, Facultade de Ciencias, Universidade de Vigo, Ourense, Spain

INTRODUCTION

As a subject of study, the atmospheric sciences encompass all the processes that occur in the atmosphere, together with its links with other systems, mainly the hydrosphere, cryosphere, lithosphere, biosphere, and outer space. As such it is an extensive discipline and the task of describing the main challenges is not an easy one, and entails a fair degree of overlap with some of the other grand challenges in the earth and environmental sciences. As a special overlapping could occur with climate sciences it is worth to remember that atmospheric processes differ from climate ones in the temporal scale, being the latter those occurring over long periods, typically higher than 30 years, but in any case long enough to produce meaningful averages. Atmospheric processes are central to configure the state of the climate but also to many of the forcings and feedbacks that determine the magnitude of climate change and its possible impacts. Additionally, there has been impressive progress of late in the atmospheric sciences in terms of the benefits provided to individuals and organizations. The flow of atmospheric "information" is of considerable importance in decisions related to health, agriculture, energy, power, and the environment. This "Grand Challenges" article focuses on the atmosphere, although the strong interaction with other parts of the earth and its environment, together with the societal implications involved, is a common theme in all the challenges described.

Over the next few years, progress in the atmospheric sciences is essential if understanding of the basic processes and their modeling is to improve; this will require genuine advances in observational, conceptual, and technological approaches. For this reason the following non-exhaustive list of 12 selected challenges includes those related to observations and data assimilation, those covered within the traditional disciplines (atmospheric physics and chemistry, atmospheric dynamics and weather forecasting), those concerned with the interactions between the atmosphere and its boundaries, and those related to the atmospheric component of climate studies.

CHALLENGE 1: DATA ASSIMILATION

The challenges in terms of data assimilation for earth observation over the next years relate to technical and general thematic aspects, as well as to the ability to take advantage of new and exciting opportunities in earth observation systems. The benefits of addressing these challenges are likely to include improvements to reanalyses, improvements to weather forecasting, an improved observational system, and an improved foundation on which the elements of climate models can be built. Among the technical challenges, five areas are most significant: (1) the assimilation of coupled data to account for links between different elements of the earth's system; examples include the coupling of the atmosphere and the ocean, of the ocean and the cryosphere, and of the atmosphere and the land; (2) assimilation of ensemble data to account for natural variability and/or to represent errors in the earth system—here, the technical effort will focus on the design of realistic ensembles; (3) performing data assimilation at increased spatial resolutions, representing the earth system at finer scales (mesoscale and finer), including theoretical developments to account for changes in balance conditions; (4) better representation of errors (random and bias) in the observations and models used in data assimilation, including the representation of forecast errors, model errors and online bias correction; (5) extension and consolidation of the joint state estimation and the inverse modeling approach in order to study biogeochemical cycles (e.g., the carbon cycle). The overarching challenge here is the consolidation and integration of the community data assimilation efforts of the meteorological and space agencies, of research and operational activities, and from *in situ* and satellite observational platforms, including all continental and global collaborations, and the effective application of these efforts toward the development of new missions in earth observation.

CHALLENGE 2: SMALL SCALE PROCESSES IN THE ATMOSPHERE

Several challenges are apparent in terms of our fundamental understanding of small scale processes and related applications, many of which are in currently being actively debated and studied. First, increased computational power allows the more detailed simulation of fluid mechanics problems, thus, even stably stratified flows are now modeled by direct numerical simulation. At the same time, these advanced computational techniques also require a new generation of parameterization schemes for numerical weather prediction (NWP) and climate modeling. At high resolutions, for example, the complex dynamics that occur in urban areas cannot be neglected and specific NWP schemes to represent these are required. At smaller grid sizes the so-called gray zone of turbulence is approached in NWP, and the impact of this must be understood and quantified. There is some room for improvement in terms of the representation of clouds and of the diurnal cycle of deep convection, and the same also applies to the physical processes that govern the stable boundary layers and the diurnal cycle, and the intermittent nature of turbulence, especially under calm conditions. In addition, higher resolutions also require more advanced techniques to allow the interpretation of the observations made. In boundary-layer meteorology, the closure of the surface energy balance and the heat budget in field observations requires further attention. Finally, the data challenges facing meteorology will also increase, due in particular to the greater availability of both professional and crowd-sourced observations (Muller et al., 2013).

CHALLENGE 3: AIR POLLUTION CHEMISTRY

The key components of a program to address the most important challenges for researchers in air pollution chemistry may be described under the following three headings: (1) *Indoor Pollution and health:* given the tendency for people to remain largely indoors for work, school, and leisure, it is important to study the impact of indoor pollution on human health as a result of indoor emissions and/or the infiltration of the external ambient air. In recent years the processes that govern indoor air quality have changed markedly as result of modifications to building regulations with the aim of better energy efficiency. There are still considerable unknowns in relation to the sources, compounds and processes that affect indoor air quality and its impact on human wellbeing. (2) *Dust and air quality:* with continuous improvements in the characteristics of vehicle emissions, the effects of aerosol pollution in urban areas can now increasingly be traced to other sources of emissions, such as the transport of natural dust and the resuspension of road dust, mainly in southern European areas with drier climates in areas affected by the transport of dust from the deserts of North Africa. An understanding of these impacts and the application of mitigation measures (for road dust resuspension) are both areas of future research. (3) *Biomass burning:* with climate change and concerns about the impact and cost of fossil fuels, biomass combustion is now commonly used for domestic heating in Europe. In many urban areas, principally in winter, domestic biomass has been found to be an important source of air pollution by particulates. Ar present some emphasis is being placed on the evaluation of the impact of biomass burning in terms of urban air quality as well as in terms of the study of the emission characteristics of biomass burning equipment and installations, as well as on the impact of the composition of biomass burned particles on human health.

CHALLENGE 4: AEROSOL-CLOUD INTERACTIONS

There is no doubt that aerosol particles are actively involved in cloud formation via the supply of cloud condensation nuclei (CCN) and ice nuclei (IN). It has been suggested that changes in aerosol concentrations will alter cloud lifetimes and precipitation efficiency, and hence affect the radiative forcing of the earth system. Great efforts have been devoted to this topic, resulting in rapid developments in terms of knowledge, methodologies, and techniques (e.g., Wang, 2013). Despite this progress, it is still difficult to draw any meaningful conclusions about the climatological effects of aerosols at regional and global scales. In contrast, aerosol-cloud interactions at molecular and microphysical scales have become more and more predictable and its modeling more deterministic. There appears to be a significant gap in our knowledge between the small-scale (molecular and microphysical) processes and the large-scale (regional/global) events in this area. We suggest that there remains a need to synthesize multi-scale results to identify clearly the problems involved and to improve the current set of tools and methodologies required to close the gap.

CHALLENGE 5: WEATHER PREDICTION

Phenomena described by fluids are complex, however, the appearance of the laws of fluid motion is deceitfully simple, equations governing these laws are non-linear, what implies multiple (and hard to understand) types of feedback effects. The atmosphere and the temporal evolution of its state does not deliver from this problem. In any case one of the flagships of the body of research on atmospheric sciences over the last few decades has been the establishment of reliable forecasting in the 2–7-day range, in view of the enormous potential economic benefits; however, such techniques still suffer from problems derived from the collection and utilization of data, which are mostly collected over the oceans. The use of new data from satellites and ground-based remote sensing could help in this regard, as could the correct maintenance of traditional data sets such as the now some what outdated global rawinsonde network. Improvements in measurements of water vapour and land surface properties are also priorities. The physical challenges continue to be the same as they were when defined more than a decade ago (National Research Council, 1998), namely: a better understanding of the nature of the interaction between atmospheric and land surface processes, the hydrological cycle, the dynamics of deep convection, the role of the tropopause in atmospheric dynamics, a fresh impetus in the development of mesoscale models and an improvement in the parameterizations used in the wave-based models of weather and climate. An example of the importance of these improvements is orographic gravity wave drag, whose parameterization in weather and climate prediction models needs to be updated given the importance of some effects shown to be important in recent research. Among these is the impact of wind shear on both the surface drag and the wave momentum flux (and its dissipation), and the drag produced by trapped lee waves, whose energy propagates, and is dissipated, downstream of their source rather than upwards. The implications of these orographic gravity waves for clear-air-turbulence (CAT), a very serious aviation hazard, have not been satisfactorily quantified. Most CAT forecast methods use empirical predictors not explicitly linked to gravity waves, but it is well known that directional shear (which is ubiquitous in nature) leads to gravity wave breaking, which may be an important source of CAT. The trapping of gravity waves in the lee of mountains or hills leads to the formation of unsteady, turbulent, closed circulations known as rotors, which are also a serious aviation hazard. Our understanding of the conditions necessary for the onset of these flow structures is incomplete, and will no doubt benefit from recent advances in mountain wave theory.

CHALLENGE 6: REMOTE SENSING FOR METEOROLOGY AND CLIMATE

Ground-based and satellite remote sensing has provided major advances in our understanding of both the weather and the climate systems, as well as the changes in these (Yang et al., 2013), by allowing the quantification of the processes and spatio-temporal states of the atmosphere, land, and oceans. The intensive use of satellite imagery in meteorology, and spatial patterns of sea level rise, provide good examples of this. The duration of the time series concerned are usually too short to allow their use for capturing long term trends of many climate variables, so one major challenge is to extend the durations of these time series. Remote sensing of the regional and global cycles of clouds and precipitation is also necessary for climate monitoring and the verification of model outputs. There are two notable challenges in atmospheric physics; the first is to design innovative studies focusing on cloud microphysics and the relationship with the physics of lightning discharge, together with all aspects related to the observation and measurement of atmospheric electricity, and the second is to develop new passive radiometer and radar studies to help us to understand the structure of clouds and precipitation with special emphasis on tropical warm rain processes, mid-latitude light precipitation, snowfall, cloud liquid and ice water content, precipitable water and water vapour profiles. One hydrometeorological challenge is to extend and improve our observations and modeling of the atmospheric and continental parts of the water cycle in order to allow its closure (e.g., mountain areas, polar regions).

CHALLENGE 7: THE ATMOSPHERIC BRANCH OF THE HYDROLOGICAL CYCLE

Among the many challenges related to the hydrological cycle, those concerned with the atmospheric transport of moisture must receive special mention because of their existence entirely within the realm of the atmospheric sciences. Here we consider the most pressing of the challenges described in the recent review of Gimeno et al. (2012). The diagnosis of moisture sources has become a major research tool in the analysis of extreme events (e.g., floods, droughts), and can be thought of as a basic tool for regional and global climatic assessments; it is therefore, necessary to check the consistency of the different approaches used to establish source-sink relationships for atmospheric water vapour. Of key importance is the improvement of our understanding of how sources of moisture affect precipitation isotopes; this is important in and of itself but it is also crucial for correctly interpreting the most prominent paleoclimatic archives including ice cores and cave sediments. A further challenge is the better understanding of the role of the transport of moisture as the main factor responsible for meteorological extremes (heavy rainfall via structures such as low level jets and atmospheric rivers, or drought via the prolonged diminished supply of water vapor from moisture source regions). In order to assess whether the moisture source regions have remained stationary in past years, it is necessary to understand the effects of the main modes of climate variability on the variability of the moisture regions, and how the transport of moisture occurs in a changing climate. These unsolved questions constitute a substantial challenge for climate scientists.

CHALLENGE 8: INTERACTION OF SCALES IN CLIMATE SIMULATION

The interaction among various spatial and temporal scales results in what we call climate. (Lorenz 1967) was among the first to emphasize the importance of scale interactions in explaining some of the key characteristics of climate observed in various regions. The non-linear character of most of these scale interactions has made them difficult to model, and as a consequence this still constitutes a source of uncertainty in climate simulations. Some empirical methods have been proposed to downscale the output from climate models but these are still somewhat controversial (Pielke and Wilby, 2011), particularly when used to interpret long term climate projections at a regional scale. The use of boundary conditions from an global model in which coupled interactions among all the major subsystems of the climate system (atmosphere, ocean, biosphere, and cryosphere) are predicted has a number of problems as the retention of large-scale climate errors in the global models, its great dependence on the lateral boundary conditions or the lack of two-way interaction between the regional and global models. The role of small scale atmospheric processes, usually in short lived phenomena, turns to be highly relevant particularly in tropical regions, where mesoscale convective systems interact with large scale circulations, and are of crucial importance in the hydrological cycle.

For example, tropical cyclones may result in very wet or dry years in some regions depending on their activity and trajectory. This element is rather difficult to simulate in climate models, but its contribution to regional climate is beyond doubt and must be better understood in order to incorporate it into climate modeling systems.

CHALLENGE 9: EXTREME EVENTS

In recent years the effects of different meteorological and climate phenomena have gained in importance in the eyes of the media and the population as a whole, partly as a consequence of extreme events such as the heatwaves in Europe (2003), Russia (2010), or USA (2011), or the deadly and extremely costly hurricanes that have hit densely populated areas in recent years, including New Orleans (Katrina, 2005) and the Metropolitan area of New York (Sandy, 2012). Likewise, prolonged periods of drought have caused severe problems for cereal producers, including in southern Australia (2002–2010), or the south-western USA, or via the increased likelihood of forest fires (Amazonia, 2005 and 2010). Some of these extreme events are closely related to the occurrence of vigorous circulation patterns such as the North Atlantic Oscillation (NAO), or to blocking and the displacement of storm tracks and the jet stream. By definition, extremes are rare in a time series, there is therefore, a pressing need, linked to the analysis of extreme events, to extend the climatic series as far as possible, and for this reason reconstructions of the past climate based on instrumental, historical and proxy data continue to be indispensable. The recent IPCC report (IPCC, 2013) shows that this growing interest in climatic extreme events must be addressed within the wider context of climate change, given that the expected changes in global, regional and even local climates are most likely to be felt through changes in the magnitude and frequency of extreme events.

CHALLENGE 10: SOLAR INFLUENCE ON CLIMATE

It has been estimated that about 8% of recent global climate change can be attributed to solar variability, but this figure must be treated with caution given that a number of aspects of solar forcing and the mechanisms coupling solar variability to the earth's climate system remain poorly understood (Gray et al., 2010). With the increasing complexity and sophistication of atmospheric and climate models, and the need for increased accuracy of the predictions made, it is important to able to include a more complete picture of solar forcing in these models. Sources of solar forcing can be divided into radiatively and particle driven components. The scientific focus for the radiatively driven forcing is currently shifting from the global to the regional responses as driven by variations in solar spectral irradiance (SSI). A number of questions remain about the nature of the variations in SSI, how these should be implemented in models, and how they will change in future solar cycles if the sun moves away from its current grand maximum of solar activity toward a new maunder minimum. The particle driven component is further divided into energetic particle precipitation (EPP) and cosmic ray (CR) effects. The EPP effect initially influences the upper stratosphere and lower thermosphere. While the chemical effects of EPP on the atmosphere are now well understood, there is a pressing need to understand further dynamical effects, as well as the potential mechanisms and magnitudes in terms of the earth's climate. The potential influence of EPP on climate is an emerging research area, and is one that is assuming a greater importance now that climate models are extending to higher altitudes that are more directly influenced by EPP. EPP provides one of the key transport pathways from the lower thermosphere down to the stratosphere and beyond, down to the troposphere via stratosphere-troposphere coupling in the polar regions. The effect of EPP could also become more pronounced in the near future as radiative forcing becomes more influenced by a move to maunder minimum types of solar activity. The CR driven component is currently considered to be the least well understood of the sources of solar forcing, although dedicated ongoing international research efforts are being made to address this question. Resent results have suggested that although CRs may stimulate aerosol nucleation, in global terms these effects are not great, and questions remain on the physical mechanisms linking CRs and aerosol nucleation.

CHALLENGE 11: URBAN WEATHER AND CLIMATE

The urban heat island (UHI) is perhaps the best known effect of the presence of cities on the local microclimate; the air temperature in a city at night can be much higher (up to 10°C or more) than in the surrounding area. Urban Climate, an emerging branch of meteorology 20 years ago, is now a mature field of research. It covers a range of topics, from fundamental theoretical studies to more applied research, having as its main goal the application of climatic knowledge to the better design of cities around the world. Micrometeorology has always been a core area of interest in urban studies because of the scales involved. Urban climatology instrumentalists have pioneered the continuous development of instrumentation and process analysis ever since the 1970s. The processes leading to the formation of the UHI (mostly physical in nature due to the 3D shape and the materials that make up the urban fabric), emerged from these early studies. Today, a number of challenges remain in relation to the measurement of this rather complex urban boundary layer. New short-range teledetection instruments are being used to gain a specialist view of the physical processes involved. Such instrumental developments will inevitably continue. Urban climate was only tackled by atmospheric modellers when the atmospheric models reached a sufficiently high resolution (a few km) to be able to represent cities explicitly. The first models representing the exchanges of energy and water between urban surfaces and the atmosphere appeared in the early 2000s (see reviews inMasson, 2006 and Martilli, 2007), and are now being used more and more in numerical weather forecasting models. The first international intercomparisons of urban models (Grimmond et al., 2010, 2011) discussed some obvious means of improvement, for example in the representation of urban vegetation. In addition, approximately 15 years later than atmospheric models, regional climate models now have spatial resolutions compatible with urban scales. This of course presents a new challenge in the proper representation of cities in climate models. Similarly, urban meteorology studies cannot be limited to physics or chemistry, but must take account of the behavior of the inhabitants. Although biometeorological studies already exist, especially in terms of levels of human comfort, the interactions between the meteorological and social worlds, both in terms of human comfort but also in terms of meteorologically dependant energy use, for example, still form one of the main challenges for urban meteorologists.

CHALLENGE 12: OZONE DEPLETION AND RECOVERY

Although stratospheric ozone concentration minima are still seen in many regions, signs of recovery are beginning to be perceived. In the Antarctic stratosphere the concentration of halocarbons peaked around the year 2000 and then began to diminish. Current projections suggest that complete recovery could occur around the year 2050. This means that one of the major challenges is to ensure the continued monitoring both of ozone and of ozone-depleting gases in order to guarantee the recovery. Improvements in the basic understanding of processes, and simulations thereof, are especially important in the context of a changing climate. Both directions must be simulated, i.e., how a changing climate will affect the ozone layer, and how the recovery of the ozone will affect weather and climate. The so-called climate-chemistry models (CCMs, Lamarque et al., 2013) appear to be of key importance in this case.

The foregoing list of challenges for the next few years in atmospheric sciences research relates only to a few of the most urgent unsolved questions and naturally remains incomplete. The challenges described herein must not be considered to be the likely principal research topics in Frontiers in Atmospheric Science; any interesting work linked to the umbrella of atmospheric science should find accommodation in the journal.

ACKNOWLEDGMENTS

Supported in part by MINECO (Spain), project TRAMO and FEDER. The team of Associated Editors of Frontiers in Atmospheric Science provides useful comments.

REFERENCES

1. Gimeno, L., Stohl, A., Trigo, R. M., Dominguez, F., Yoshimura, K., Yu, L., et al. (2012). Oceanic and terrestrial sources of continental precipitation. *Rev. Geophys.* 50:RG4003. doi: 10.1029/2012RG000389

2. Gray, L. J., Beer, J., Geller, M., Haigh, J. D., Lockwood, M., Matthes, K., et al. (2010). Solar influences on climate. *Rev. Geophys.* 48:RG4001. doi: 10.1029/2009RG000282

3. Grimmond, C. S. B., Blackett, M., Best, M. J., Barlow, J., Baik, J-J., Belcher, S. E., et al. (2010). The international urban energy balance models comparison project: first results from phase 1. *J. Appl. Meteorol. Climatol.* 49, 1268–1292. doi: 10.1175/2010JAMC2354.1

4. Grimmond, C. S. B., Blackett, M., Best, M. J., Barlow, J., Baik, J-J., Belcher, S. E., et al. (2011). Initial results from Phase 2 of the international urban energy balance model comparison. *Int. J. Climatol.* 31, 244–272. doi: 10.1002/joc.2227

5. IPCC. (2013). *Climate Change: The Assessment Report of the Intergovernmental Panel on Climate Change*. Cambridge, UK: Cambridge University Press.

6. Lamarque, J.-F., Shindell, D. T., Josse, B., Young, P. J., Cionni, I., Eyring, V., et al. (2013). The atmospheric chemistry and climate model intercomparison project (ACCMIP): overview and description of models, simulations and climate diagnostics. *Geosci. Model. Dev.* 6, 179–206. doi: 10.5194/gmd-6-179-2013

7. Lorenz, E. N. (1967). *The Nature and Theory of the Atmosphere*. Geneva: WMO, 161.

8. Martilli, A. (2007). Current research and future challenges in urban mesoscale modelling. *Int. J. Climatol.* 27, 1909–1918. doi: 10.1002/joc.1620

9. Masson, V. (2006). Urban surface modeling and the meso-scale impact of cities. *Theor. Appl. Climatol.* 84, 35–45. doi: 10.1007/s00704-005-0142-3

10. Muller, C. L., Chapman, L., Grimmond, C. S. B., Young, D. T., and Cai, X. (2013). Sensors and the city: a review of urban meteorological networks. *Int. J. Climatol.* 33, 1585–1600. doi: 10.1002/joc.3678

11. National Research Council. (1998). *The Atmospheric Sciences: Entering the Twenty-First Century*. Washington, DC: The National Academies Press.

12. Pielke, R. A. Sr., and Wilby, R. L. (2011). Regional climate downscaling—what's the point? *EOS* 93, 52–53. doi: 10.1029/2012EO050008

13. Wang, C. (2013). Impact of anthropogenic absorbing aerosols on clouds and precipitation: a review of recent progresses.*Atmos. Res.* 122, 237–249. doi: 10.1016/j.atmosres.2012.11.005

14. Yang, J., Gong, P., Fu, R., Zhang, M., Chen, J., Liang, S., et al. (2013). The role of satellite remote sensing in climate change studies. *Nat. Clim. Change* 3, 875–883. doi: 10.1038/nclimate1908

The Influence of Climate Factors, Meteorological Conditions, and Boundary-layer Structure on Severe Haze Pollution in the Beijing-tianjin-hebei Region during January 2013

Lili Wang[1], Nan Zhang[2], Zirui Liu[1], Yang Sun[1], Dongsheng Ji[1], and Yuesi Wang[1]

[1]LAPC, Institute of Atmospheric Physics, Chinese Academy of Sciences, Beijing 100029, China

[2]Hebei Province Meteorological Observatory, Shijiazhuang 050022, China

ABSTRACT

The air-pollution episodes in China in January 2013 were the most hazardous in the Beijing-Tianjin-Hebei (BTH) region. $PM_{2.5}$, AOD, and long-term visibility data, along with various climate and meteorological factors and the boundary-layer structure, were used to investigate the cause of the heavy-haze pollution events in January 2013. The result suggests that unfavorable diffusion conditions (weak surface winds and high humidity) and high primary-pollutant emissions have induced heavy-haze pollution in the BTH region over the past two decades. A sudden stratospheric warming (SSW), weak East Asian winter monsoon, a weak Siberian High, weak meridional circulation, southerly wind anomalies in the lower troposphere, and abnormally weak surface winds and high humidity were responsible for the severe haze pollution events, rather than an abrupt increase in emissions. Heavy/severe haze pollution is associated with orographic wind convergence zones along the Taihang and Yanshan Mountains, slight winds (1.7~2.1 m/s), and high humidity (70%~90%), which limits the diffusion of pollutants and facilitates the hygroscopic growth of aerosols. Recirculation and regional transport, along with the poorest diffusion conditions and favorable conditions for hygroscopic growth of aerosols and secondary transformation under the high emission, led to explosive growth and the record high hourly average concentration of $PM_{2.5}$ in Beijing.

INTRODUCTION

Haze pollution in China has increased over the past three decades, particularly in city clusters, as a result of the rapidly developing economy, expanding anthropogenic activities and urbanization [1, 2]. The Beijing-Tianjin-Hebei (BTH) city cluster located in the North China Plain, which includes two megacities (Beijing and Tianjin) and Hebei province, has one of the largest populations (Figure 1(a)) and the third largest gross domestic product in China [3–5]. In recent years, large-scale regional haze pollution characterized by high concentrations of $PM_{2.5}$ (particulate matter with aerodynamic diameters ≤ 2.5 µm) has frequently occurred [6–8], and the levels of $PM_{2.5}$ continue to frequently exceed the national ambient air quality standards of China [9]; as a result, serious environmental, climatic, and health problems are rampant [10, 11].

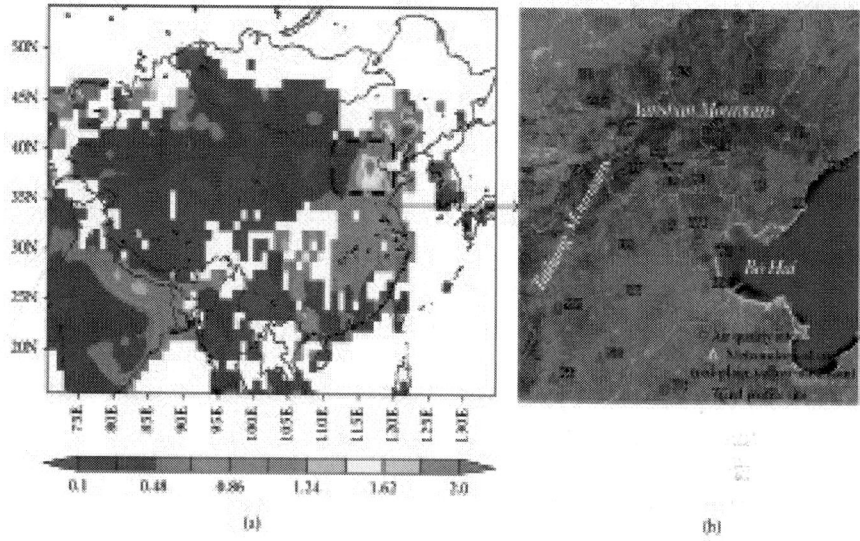

(a)   (b)

Figure 1: (a) Monthly mean deep-blue AOD at 550 nm wavelength in January 2013 over China derived from Aqua MODIS 1° × 1° level-3 daily products. (b) Topographic map and the locations of meteorological and air quality measurement sites over the BTH region.

A relatively high level of $PM_{2.5}$ is usually associated with the close proximity of high precursor emissions [12]. In addition, weather/climate features play a significant role in air quality through a complex combination of processes and influences, such as emissions, transport, chemical transformations, and wet or dry removal processes. Severe pollution episodes in the urban environment mainly result from unfavorable air dispersion related to weather conditions [13, 14]. Researchers have investigated that the effects of climate change on haze/$PM_{2.5}$ pollution and revealed that a significant contribution came from the weakening monsoon circulation in past decades, trapping more pollutants over Eastern China [15, 16]. On the synoptic scale, high pressure and successive low pressure systems are associated with the formation of air pollution, and pressure systems with strong gradients lead to the decrease of the air-pollution index (API) over North China [17]. Moreover, the BTH region is surrounded by mountains in the north and west directions and by Bohai Sea in the east direction (Figure 1(b)). Unique local topographic factor, together with stagnant weather with weak wind and temperature inversion, and regional transported contribution by south wind can induce heavy air pollution [18–20]. Due to the combustion of coal for heating and the unfavorable dispersion attributable to geographical and meteorological conditions, haze pollution in winter over the BTH region is more serious [21,22]. The governments, especially the Beijing municipal government, have produced a series of emission reduction measures to control air pollution [4, 23], such as changes of industrial structure and improved energy efficiency, the use of clean energy and preferred use of clean coal and gases, vehicle limitation measures, and implementation of advanced environmental standards. However, the pollution levels are still very high all the year round. Hazardous air-pollution episodes occurred in January 2013 and affected the BTH region of China the most (Figure 1(a)); thus, this event has attracted extensive attention [6, 19, 24].

Scientists have investigated the effect of regional transport and local accumulation during stagnant weather and the secondary formation of particulate on severe regional $PM_{2.5}$ pollution events in January 2013 [6, 19,25, 26]. Notably, the emission source did not increase rapidly in January 2013. Zheng et al. elaborated that [27] the changes of monthly averaged emissions in January over the Beijing-Tianjin-Hebei region were increased in rates, that is, 2.1%, 1.5%, and 2.5% for primary PM2.5, SO_2, and NO_x, respectively, compared with emissions in January 2012, but they were still not significant compared to the changes in pollutant concentrations and the weather/climate elements was principally responsible for these heavy pollution episodes. Zhang et al. [28] pointed out that the weakened surface winds, the anomalous southerly winds in the middle and lower troposphere, and the anomalous inversion were associated with the strong haze/fog event in Eastern China by diagnosing both its atmospheric background fields and daily evolution in January 2013. However, few studies have explored the characteristics and formation of heavy $PM_{2.5}$ pollution in regard to climate factors, meteorological conditions, and boundary-layer structure over the BTH region; these factors will be important for implementing effective control measures of air pollution on the regional scale.

In this study, $PM_{2.5}$, AOD, and long-term visibility data, along with various climate and meteorological factors and the boundary-layer structure, were used to investigate the cause of serious haze pollution events in January 2013 and to determine the relevant meteorological features during the most hazardous $PM_{2.5}$ pollution episode (10–14 January, 2013).

MEASUREMENT SITES AND DATA DESCRIPTIONS

Sites and $PM_{2.5}$ Measurements

The locations of the 22 meteorological observation sites and 7 air quality sites selected for this study over the BTH region are displayed in Figure 1(b). Detailed information on the air quality sites is given in Table 1. Fourteen meteorological sites are located in the plains area, and 8 sites are located in the mountainous regions. Seven air quality sites (including 4 urban sites: Beijing Tower (BJT), Shuangqing Lu (SQL), Tianjin (TJ), and Shijiazhuang (SJZ); 1 suburban site: Xianghe (XH); 1 rural site: Yucheng (YC); and 1 regional background site: Xinglong (XL)) are local in the southern plain, where the air pollution is most severe. Air quality monitoring was conducted via an observation network established by the Institute of Atmospheric Physics (IAP), Chinese Academy of Sciences (CAS). The online measurement of the mass concentration of $PM_{2.5}$ or PM_{10} was obtained by a Tapered Element Oscillating Microbalance (Model 1400A, R&P), and data were collected every five minutes. The instrument's measuring principle coincides with that of the US Environmental Protection Agency (EPA) (http://www.epa.gov/ttnamti1/inorg.html). The hourly average concentrations were calculated with the 5 min data; and daily average concentrations were acquired based on 1 h data. Monthly mean deep-blue AOD at 550 nm wavelength in January 2013 over China derived from Aqua MODIS 1° × 1° level-3 daily products was shown in Figure 1(a), which revealed that the BTJ region was the most pollution area in China.

Table 1: Locations of air quality sites over the BTH region

Station name	Station type	Latitude (°N)	Longitude (°E)	Altitude (m)
Beijing Tower (BJT)	Megacity	116.37	39.97	44
Shuangqing Lu (SQL)	Megacity	116.33	39.99	45
Tianjin (TJ)	Megacity	117.21	39.08	20
Shijiazhuang (SJZ)	Megacity	114.53	38.03	70
Xianghe (XH)	Suburban	116.96	39.75	9
Yucheng (YC)	Country	116.37	36.67	37
Xinglong (XL)	Regional background	117.58	40.39	960

The air-pollution index (AQI) is a quantitative measure describing air-pollution levels in China based on the conversion of air-pollution data, mainly PM_{10}, $PM_{2.5}$, SO_2, NO_2, CO, and O_3 concentrations, into a single value ranging from 0 to 500. The API is divided into six ranks representing different air quality levels with respect to their impacts on human health. The first and second air quality levels (I: AQI = 0–50 and II: AQI = 51–100) represent good air quality and no human-health risk. The third and fourth levels (III: AQI = 101–150 and IV: AQI = 151–200) indicate light pollution affecting human health to some degree. The fifth and sixth levels (V: AQI = 201–300 and VI: AQI > 300) denote heavy pollution and serious effects on human health. The AQI of the city is defined by the maximum subindex among all the six pollutants. In January 2013, the major pollutant was $PM_{2.5}$; thus, air quality levels were divided by air quality index (AQI) according to $PM_{2.5}$ concentrations. The corresponding concentration at the breakpoint in different air quality levels is 35, 75, 115, 150, and 250 $\mu g/m^3$, respectively.

Meteorological Data

The meteorological data (daily average and monthly average) at the plains sites and mountainous sites, the visibility and relative humidity (RH) at 14:00 LST at six plains sites since 1980 (Figures 1(b) and 2), and synoptic maps were obtained from the National Meteorological Information Center and China Meteorological Administration. The visibility and RH at 14:00 LST from 1960 to 1979 were from NOAA's National Climatic Data Center (NCDC); data was unavailable during 1965–1972. In this work, haze events were defined as days with the visibility (at 14:00 LST) <10 km and the RH < 90%, excluding days with fog, precipitation, dust storms, smoke, snow storms, and so forth. This definition is more appropriate for quantifying haze days compared with the definition using daily mean meteorological data in the BTH area [2, 21, 29]. The hourly meteorological data, wind direction, wind speed, and RH are measured with automatic weather stations (AWSs) in BJ, TJ, SJZ, XH, YC, and XL. The wind parameters are observed at a 10 m height, and the moisture information is collected at 1.5 m height. In addition, 370 AWSs with wind data and humidity information within 36°–42.5°N and 113.5°–120°E are used to obtain the gridded hourly surface-wind field and relative humidity using the Grid Analysis and Display System (GrADS) software. Gridded surface RH and wind data from the National Centers for Environmental Prediction (NCEP)/ National Center for Atmospheric Research (NCAR) final global forecast system (FNL) reanalysis dataset (1° horizontal resolution) were also used to reveal the characteristic wind field and moisture distribution that have led to haze events in the BTH region. The wind directions in this paper are referred to as the directions the winds are coming from. In the Beijing urban area and Tianjin suburban area, the wind profiles with a temporal resolution of 6 min were observed with a boundary-layer wind-profile Lidar (Figure 1(b)). The radiosonde data at the Beijing meteorological station (54511) were used to analyze the vertical structure of the temperature and humidity.

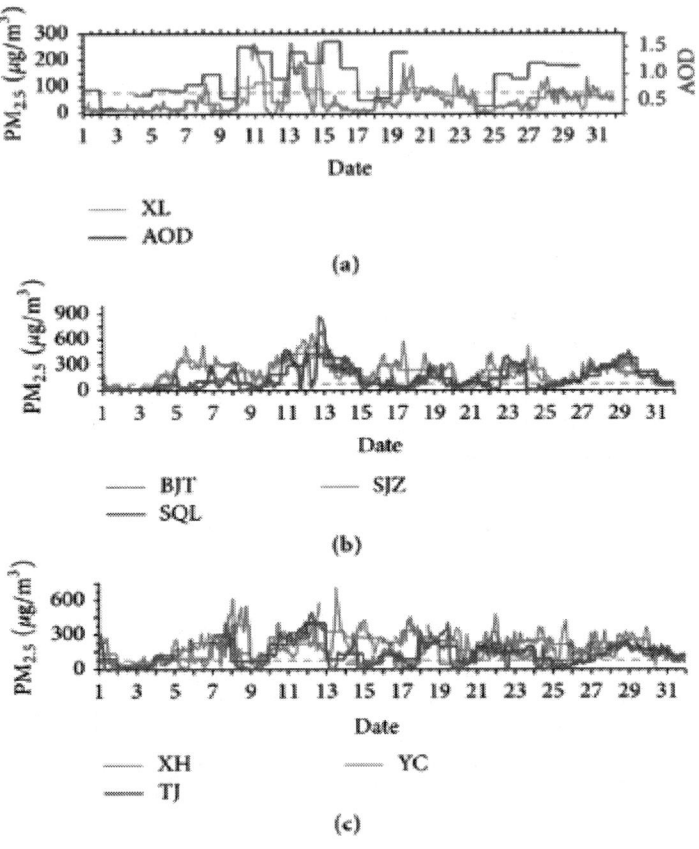

Figure 2: Variation of hourly average (sawtooth lines) and daily average (step lines) concentrations of $PM_{2.5}$ in different sites and regional average AOD in January 2013 over the BTH region. Dashed lines show the Grade II National Ambient Air Quality Standard.

Zonal circulation index for Asia (IZ), meridional circulation index for Asia (IM), East Asian winter monsoon (EAWM) intensity index, Siberian High (SH) intensity index, 1000 hPa wind anomaly, and integrated moisture flux anomaly were provided by National Climate Center (http://cmdp.ncc.cma.gov.cn/cn/index.htm). The anomalies are calculated from their departures from 1981–2010 mean values. NCEP GDAS-CPC zonal temperature anomaly downloaded from http://www.cpc.ncep.noaa.gov/products/stratosphere/strat-trop/ was used to analyze a sudden stratospheric warming (SSW).

Satellite Data

Moderate resolution imaging spectroradiometer (MODIS) onboard the Terra and Aqua satellites provides a high spatial resolution and near daily global coverage aerosol optical depth (AOD) data [30, 31]. The MODIS AOD products (MOD04/MYD04) over land comprise two datasets: dark-target AOD data and deep-blue AOD data (http://modis-atmos.gsfc.nasa.gov/MOD04_L2/index.html). The former is applicable to dark surfaces, and the latter is applicable to bright desert surfaces [32–35]; these datasets have been released as MODIS aerosol products, Collection 051, with spatial resolutions of 10 km × 10 km or 1° × 1° (http://modis-atmos.gsfc.nasa.gov/MOD04_L2/atbd.html). Because less vegetation cover exists in winter over the BTH region, the deep-blue AOD data at 550 nm from MYD04_L3 C051 were used. Aqua MODIS 1° × 1° level-3 daily products were averaged over the region 35°–43° N and 112°–120° E in January from 2004 to 2013.

RESULTS AND DISCUSSION

Distributions of PM$_{2.5}$ and AOD in the BTH Region

Variations in the hourly average and daily average concentrations of PM$_{2.5}$ at different sites and the regional average AOD in January 2013 over the BTH region are described in Figure 2. The most polluted site was SJZ, with a monthly average value of ~214 µg/m^3; and the second most polluted site was YC, with a value of ~203 µg/m^3. BJT, SQL, TJ, and XH were moderately polluted, with values of ~132–143 µg/m^3. The least polluted site was XL, with a value of ~49 µg/m^3. The YC site had a very high PM$_{2.5}$ level, although it was a rural site, which indicated the influence of regional transport. Overall, the variation of PM$_{2.5}$ concentrations at the different sites showed regional characteristics and periodic changes. Except XL (a regional background site), the PM$_{2.5}$ concentrations far exceeded the Grade II National Ambient Air Quality Standard (75 µg/m^3, GB3095-2012). The frequencies of the regional air quality at levels I and II, III and IV, and V and VI (the regional air quality level was based on the level occurring at a minimum of 4 sites) were 3 days, 13 days, and 15 days, respectively, and the corresponding regional mean $PM_{2.5}$ concentrations were 36 ± 33 µg/m³, 115 ± 83 µg/m³, and 224 ± 88 µg/m³, respectively. The regional average AOD was also very high, with a range from 0.35 to 1.5. The regional background site (XL) had 6 days that exceeded the Grade II standard. Two consecutive and regionally severe haze episodes were significant (10–14 and 26–30) but with different characteristics, that is, "explosive growth" and "sustained growth" of $PM_{2.5}$, respectively. However, the highest regional mean concentration ($PM_{2.5}$ of ~250 µg/m³ and an AOD of ~1.3) was observed during 10–14; meanwhile, the highest hourly concentration (881 µg/m³) was found at the SQL site in Beijing.

Effects of Climatic Factors on Haze Pollution

Several datasets were used to analyze the effects of climatic factors on haze pollution over the BTH region (Figure 3). Interannual variations were calculated for precipitation, wind speed, temperature, relative humidity, and sunshine hours' anomalies in the plains and mountainous sites. Temperature, depression of the dew point, and wind speed at 850 hPa level anomalies at BJ and XT were analyzed. IZ and IM for Asia were also displayed. Additionally, the occurrences of haze at five urban sites and a suburban site, the mean visibility for all days and haze days, and the mean AOD over the BTH region were computed for January. The precipitation presented an irregular fluctuation. The change in the averaged surface wind speed and sunshine hours over all the stations, particularly over the plains stations, showed a rapid decrease from 1960 to around 1990, whereas the wind speed remained constant in the plains sites and slightly increased in the mountainous sites; the sunshine hours exhibited a sustained decline in the plains sites and a steady oscillation in the mountain sites from 1990 to 2013; this trend is consistent with that found in many areas in China [36, 37]. Therefore, increasingly poor dispersion conditions superimposed with high aerosol loadings contributed to regional dimming. The temperature and RH increased over time as a result of industrial activities and urbanization. High moisture facilitates the hygroscopic growth of aerosols, which increases aerosol mass concentrations and extinction; these conditions lead to low-visibility events. However, the trends in the temperature, moisture, and wind speed at the 850 hPa level were not significant. IZ and IM for Asia changed dramatically over different years, and the amplitude of the variation increased. A weak negative correlation existed between the mean visibility and IZ, but a significantly positive correlation existed between the mean visibility and IM; thus good dispersion conditions resulted in high visibility when IM was high. The meridional circulation was strong, which facilitates ventilation due to the frequent exchange of southern and northern air masses.

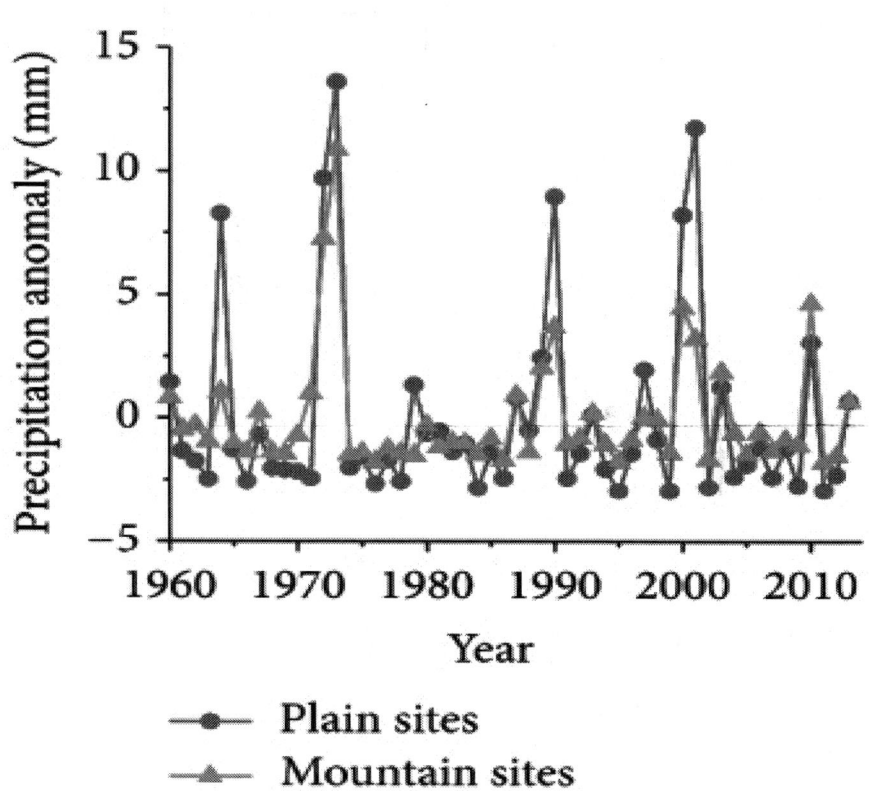

(a)

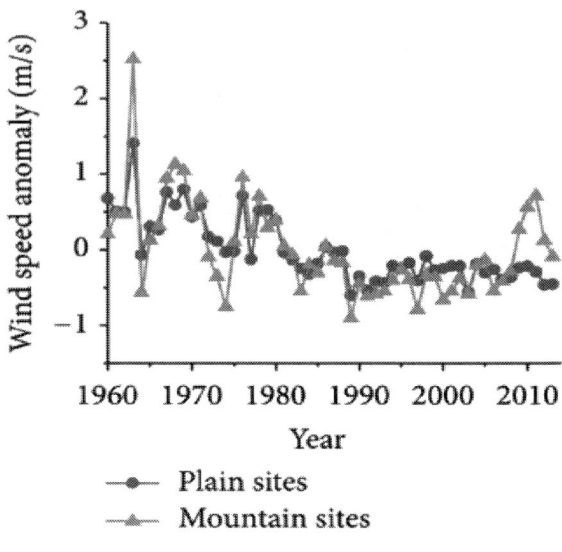

(b)

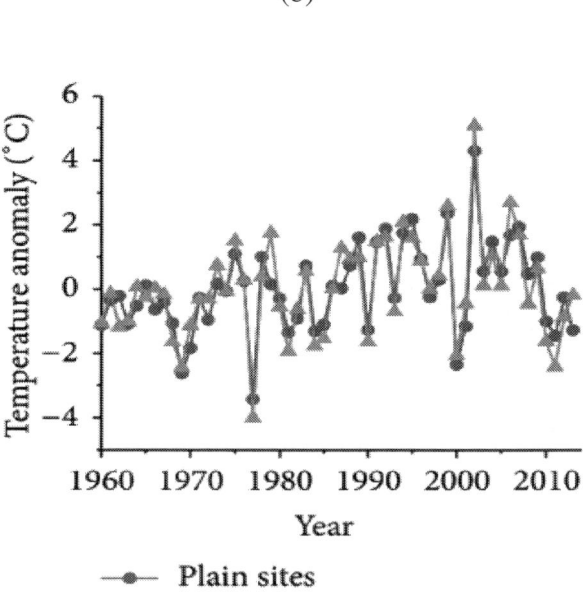

(c)

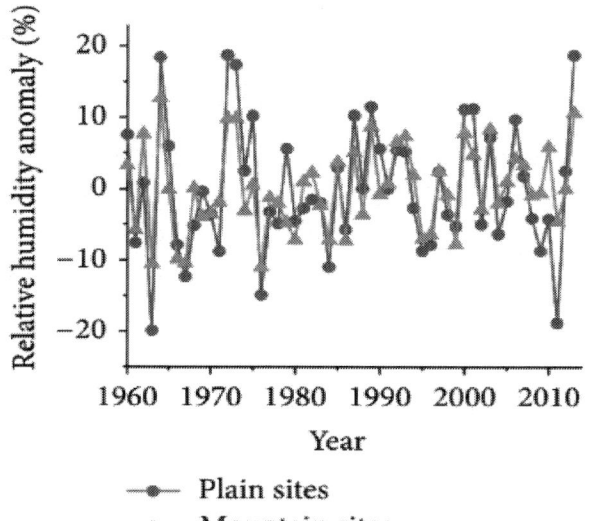

(d)

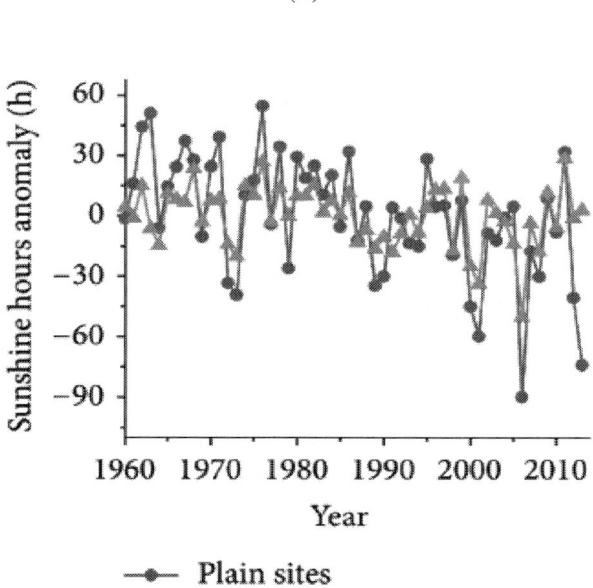

(e)

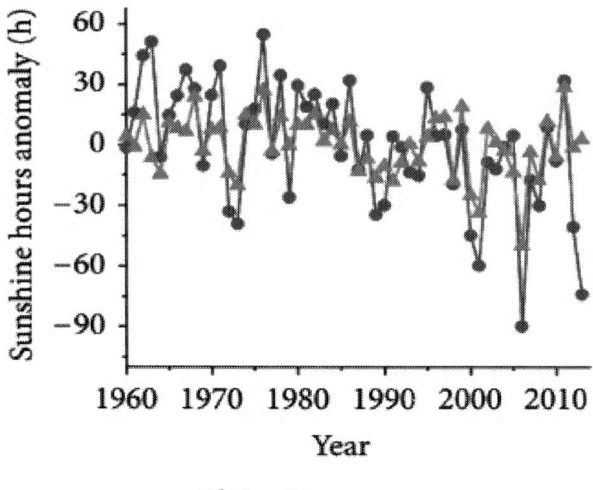

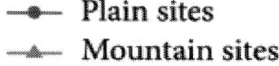

(f)

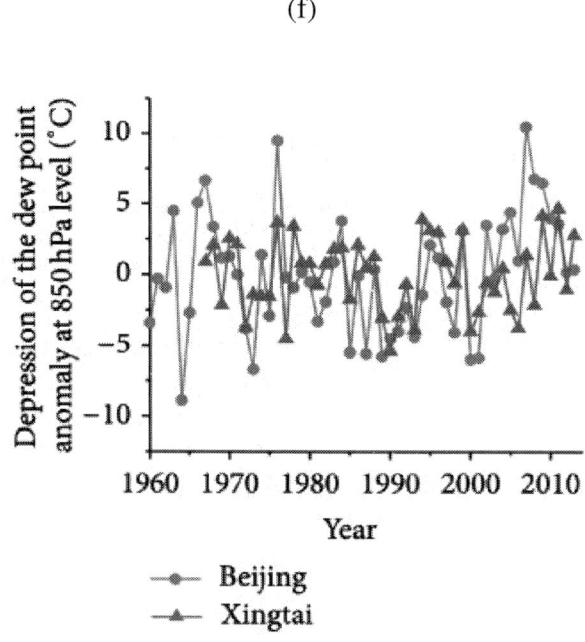

(g)

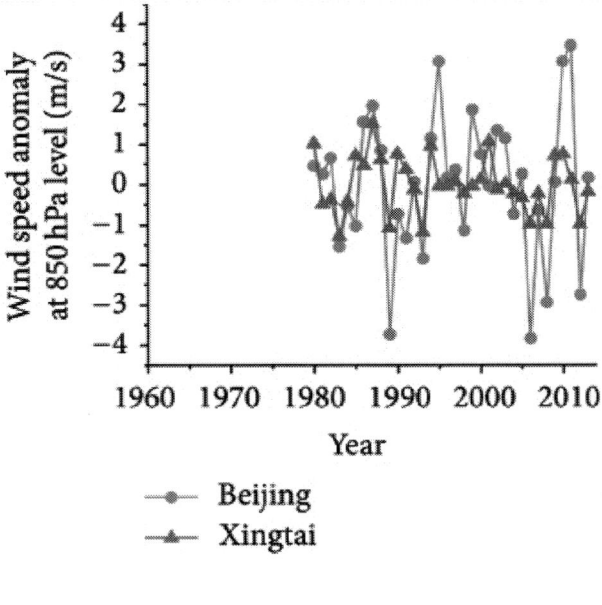

(h)

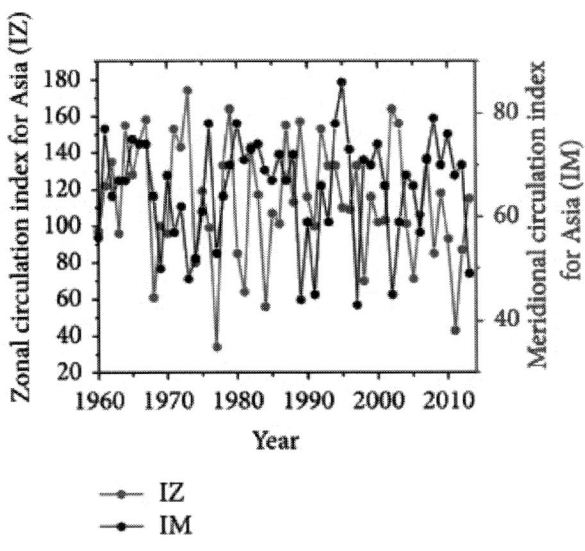

(i)

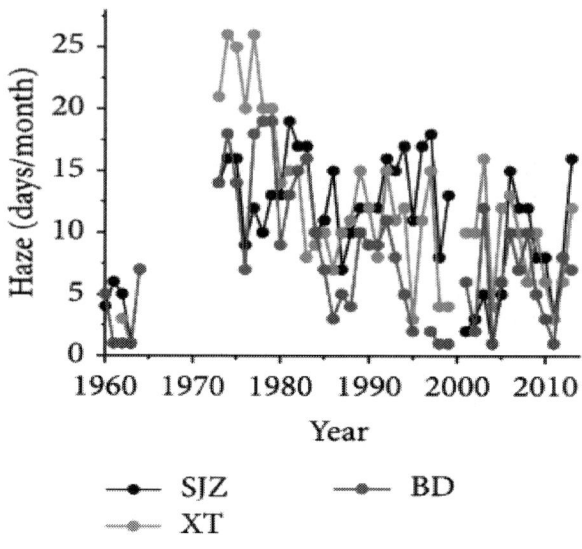

(j)

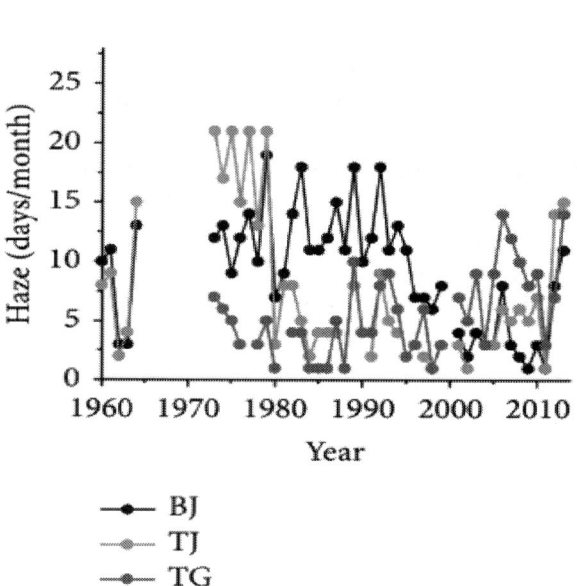

(k)

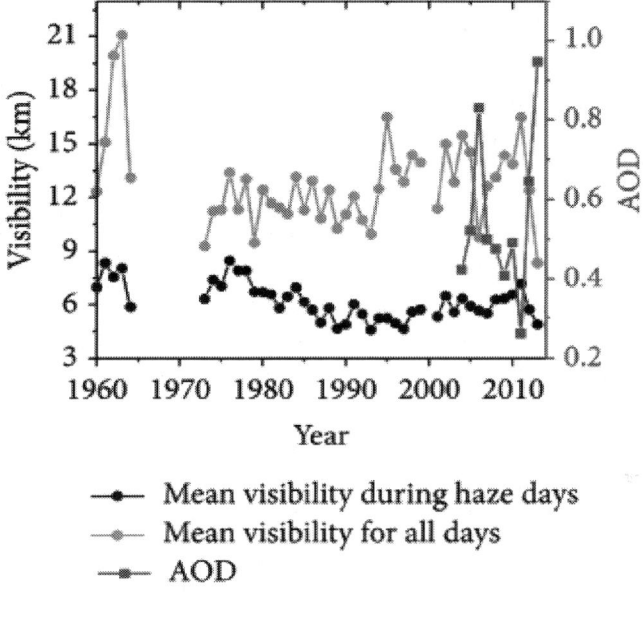

(l)

Figure 3: Interannual variations of (a)~(e) precipitation, wind speed, temperature, relative humidity, and sunshine hours' anomalies for the plains and mountainous sites, respectively; (f)~(h) temperature, depression of the dew point, and wind speed at 850 hPa level anomalies at Beijing (BJ) and Xingtai (XT), respectively; (i) IZ and IM; (j)~(k) the occurrences of haze at five urban sites (BJ, TJ, SJZ, BD, and XT) and a suburban site (TG); (l) mean visibility for all days and haze days and mean AOD over the BTH region in January. The climatology is calculated based on the mean in January from 1960 to 2013.

On average, low-visibility and haze days at the urban sites were less frequent before the early 1970s; then the frequency increased and remained a constant high value before 1998 due to the rapid economic development starting in the late 1970s. Afterward, a sudden decrease in the frequency occurred, while the amplitude of the variation changed drastically. At the suburban site, an increase in haze days occurred from the early 2000s (when this region became a special economic zone) to the present, indicating that the timing of economic development plays an important role in haze pollution. Surprisingly, the interannual variation in the mean visibility during the haze days showed a significant decrease, which indicated that the primary type of haze changed from slight haze (5–10 km) to severe or serious haze (<5 km) since the beginning of the 1990s. The results suggest that unfavorable diffusion conditions (weak surface winds and high surface humidity) combined with high primary-pollutant emissions [2] have induced very heavy-haze pollution in the BTH region over the past two decades.

Under the influences of abnormally weak wind, high humidity near the ground, high upper-air temperature, low surface temperatures, weak meridional circulations, haze pollution levels, and AOD loadings were very strong in the years 2006 and 2013, whereas the air quality in 2011 was rather good due to the opposite meteorological conditions. In addition, a sudden stratospheric warming (SSW) (Figure 4) kept the stratosphere anomalously warm in 2006 and 2013 compared with the zonal temperature anomaly distribution in 2011. This phenomenon can contribute to the reversed direction of the westerly winds of the polar vortex; these conditions weaken the polar vortex. In addition, the increasing temperature in high-latitude regions may influence the circulation and contribute to a weaker pressure gradient force. As shown in Figure 5, during the course of its rotation and collapse, a weakened polar vortex in northern Asia was associated with a weak EAWM intensity index and SH intensity index; the monsoon occasionally transported weak cold air to the eastern regions of China. When cold air transport was weak, Eastern China was under the influence of southerly wind anomalies in the lower troposphere (Figure 6(a)) and abnormal water vapor divergence (less precipitation) (Figure 6(b)), which resulted in reduced wind velocity and increased air pollution.

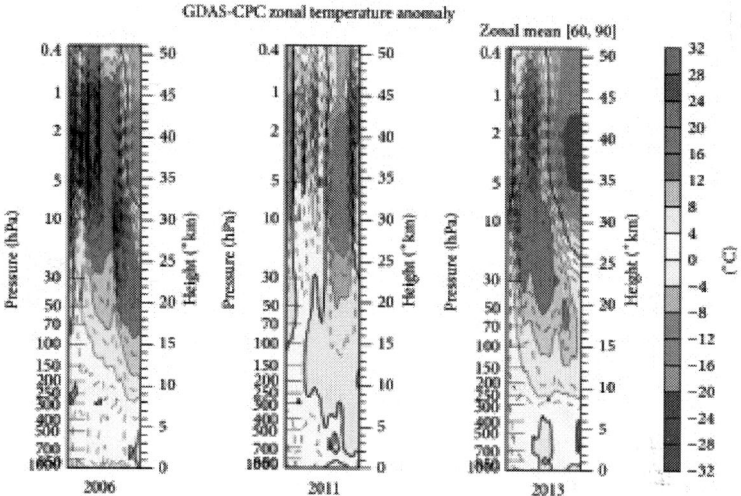

Figure 4: NCEP GDAS-CPC zonal mean time series of temperature for January of 2006, 2011, and 2013.

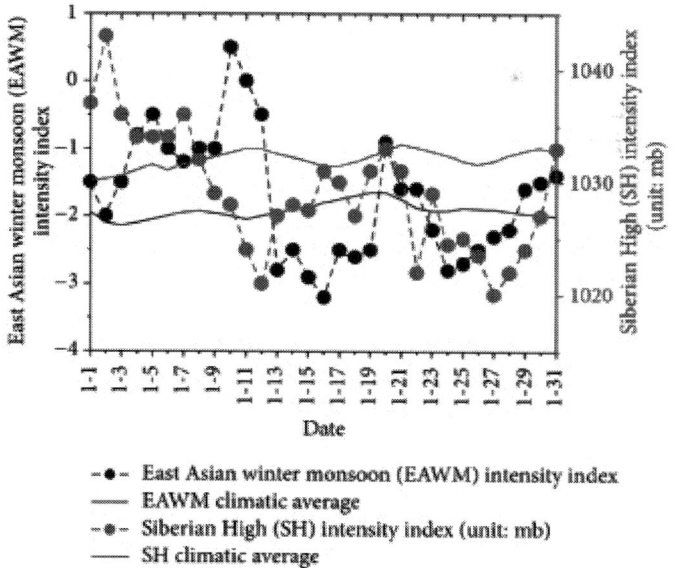

Figure 5: Time series of EAWM intensity index and SH intensity index, along with their climatic averages.

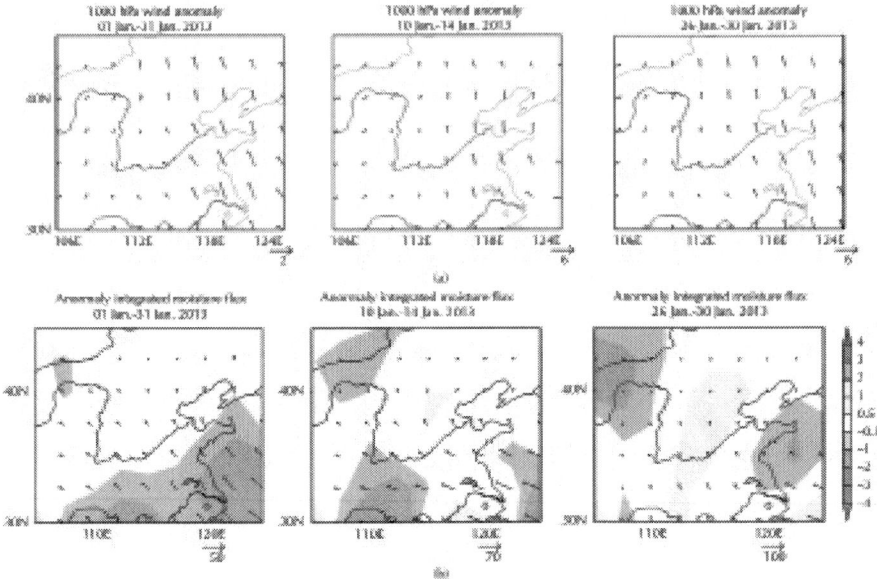

Figure 6: (a) 1000 hPa wind anomaly (vectors; m/s) (b) integrated moisture flux from 1000 to 300 hPa (vectors; kg/(s·m)) with convergence (<0) and divergence (>0) (shadings; 10^{-5} kg/(s·m²)) for different periods during January 2013 using daily and monthly mean NCEP/NCAR reanalysis data with a 2.5° resolution.

Influence of Meteorological Conditions on PM$_{2.5}$ Concentrations in January 2013

The concentration level of PM$_{2.5}$ is dependent on wind and relative humidity; wind is related to dispersion and the transport of air pollution, while RH is associated with the hygroscopicity and scattering of particles. The distributions of the average surface RH and wind field for different periods in January 2013 are depicted in Figure 7. Wind speeds in the plains area of the BTH region are weaker over North China, and the lowest values are found in the cities near the Yanshan and Taihang Mountains because of the orography and surface roughness in large cities (densely distributed high buildings). A weak southerly wind prevailed over the piedmont plain in the BTH region in January 2013. Moreover, an orographic wind convergence zone was set up along the plain-mountain transition area along the Taihang Mountains, where a concentrated contamination zone was present. Therefore, the highest concentration in SJZ was partly attributed to the convergence zone. In addition, the wind convergence zones were located in the north along the Yanshan Mountains during the two heavy-haze pollution episodes. These two orographic wind convergence zones resulted in the pollution accumulation in the piedmont plain and restrained the diffusion of pollutions; as a result, severe regional haze pollution developed. In addition, haze events are strongly influenced by the ambient RH. The RH values over the plains region of the BTH area were 65%~80% in January, 70%~90% on 10–14 January, and 80%~95% on 26–30 January. Thus, an increasing number of haze events were possibly caused by the hygroscopic growth of aerosols. Because the decreased wind speed and weakened southerly winds resulted in more stable atmospheric conditions and weaker dispersion, more effort should be paid to control emissions and prevent haze events.

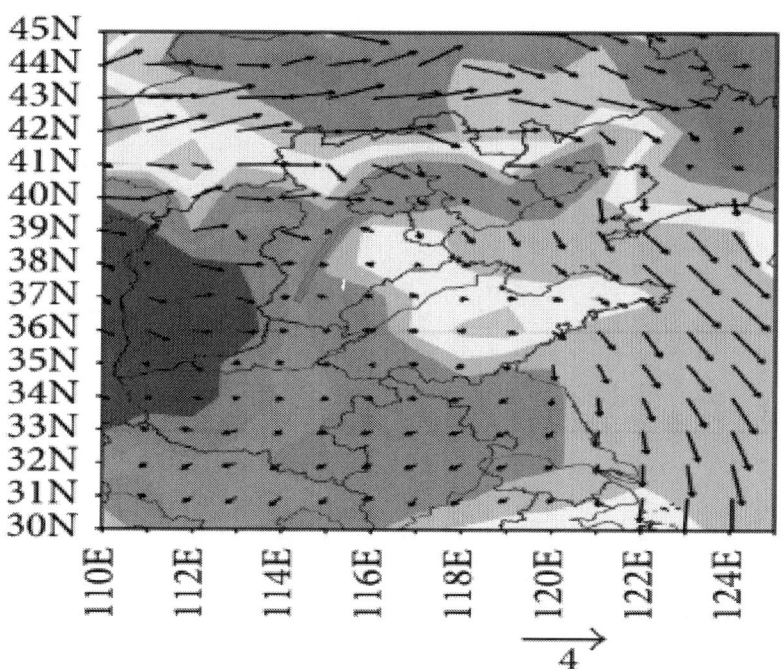

(a)

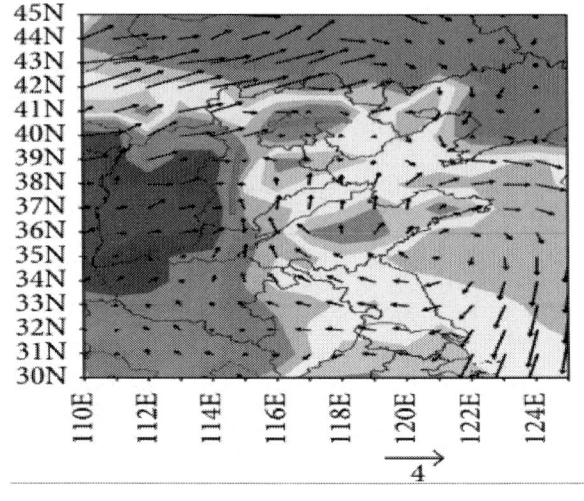

(b)

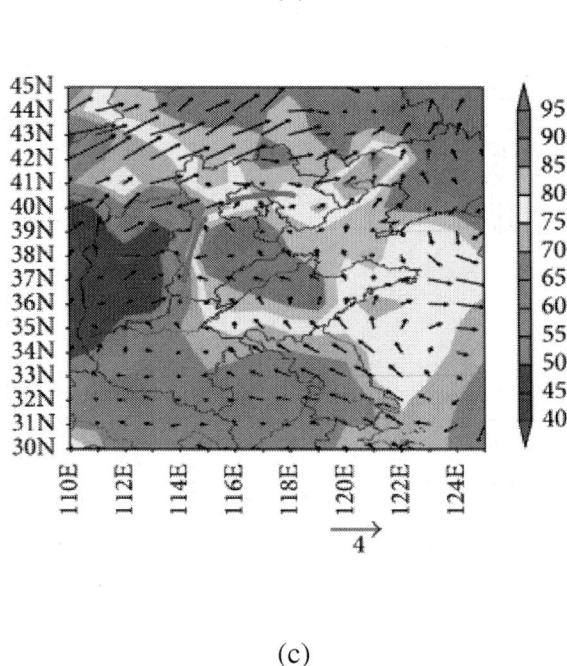

(c)

Figure 7: The distributions of the average surface RH (shaded) and wind field (black arrows) from NCEP/NCAR FNL reanalysis data for different periods in January 2013: (a) 1–31, (b) 10–14, and (c) 26–30. Red lines represent the orographic convergence line.

Meteorological factors in January 2013 for different air quality levels over the BTH region are summarized in Table 2. Compared with good air quality levels (I and II), the characteristics of the surface meteorological variables during moderate and severe haze pollution episodes in the BTH region were as follows: weaker pressure, higher temperature, particularly in the plains region, higher relative humidity, weak wind (less than 1.7~2.1 m/s), fewer sunshine hours, and lower visibility. Wind roses, the relationships between the hourly average $PM_{2.5}$ concentration and wind speed, and wind direction in January at different sites are presented in Figure 8. The lowest wind speed was found in SJZ, which is consistent with the result in Figure 7. The wind speeds associated with southeast to southwest winds were very low at most of the sites, except at YC. The northwest winds were strong in BJ and XL but weak at the remaining sites. Southwest wind, northeast wind, and calm winds were dominant. Higher $PM_{2.5}$ values were associated with weak winds (less than 3 m/s), except at XL and YC sites. At XL site, higher $PM_{2.5}$ values were associated with strong southwest, south, and north winds; and the $PM_{2.5}$ concentration at the YC site increased with the elevated wind speed, except for the northeast and southeast winds, which indicated the influence of the regional transport as the surrounding regions have high emissions.

Table 2: Summary of meteorological factors in January 2013 in different air quality levels over the BTH region

	Air quality levels		Pressure (hPa)	Wind speed (m/s)	Temperature (°C)	RH (%)	Sunshine hour (h)	Visibility (km)
All sites	I&II	Mean	1002.9	3.4	−11.8	41.5	7.9	
		SD	46.8	2.3	5.4	11.5	0.6	
	III&IV	Mean	997.5	2.1	−7.4	64.8	5.6	
		SD	45.7	1.2	4.7	16.6	3.3	
	V&VI	Mean	994.7	1.7	−5.7	76	3.1	
		SD	45.4	0.9	3.4	13.7	3.5	

Plain sites	I&II	Mean	1031.2	3.1	−9	40	8	20.3
		SD	9.6	2.3	3.2	11	0.7	6.1
	III&IV	Mean	1025.7	1.9	−5.5	68.5	5.1	10.8
		SD	5.6	1.1	3.1	16.3	3.3	4.9
	V&VI	Mean	1022.8	1.8	−4.2	82.1	1.4	4.3
		SD	4.7	0.9	2	11	2.5	3.3

SD represents standard deviation

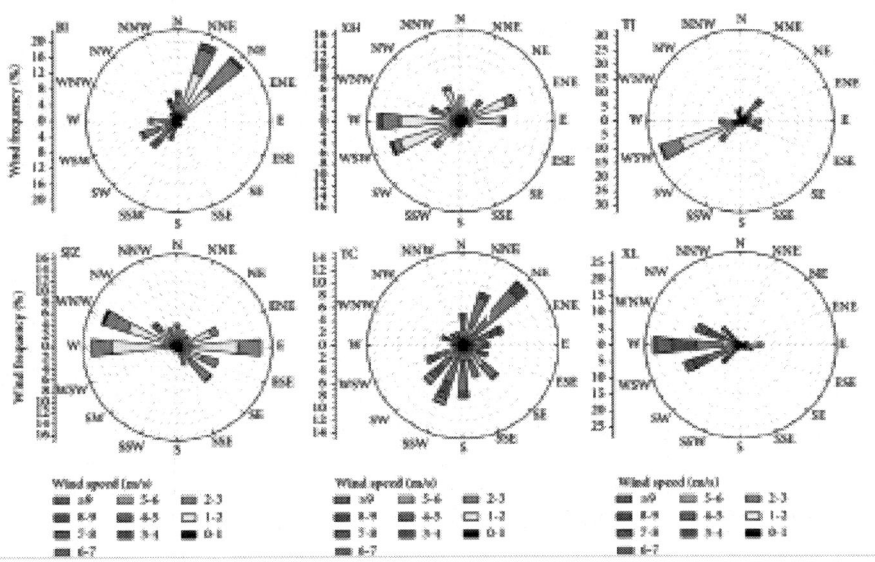

(a)

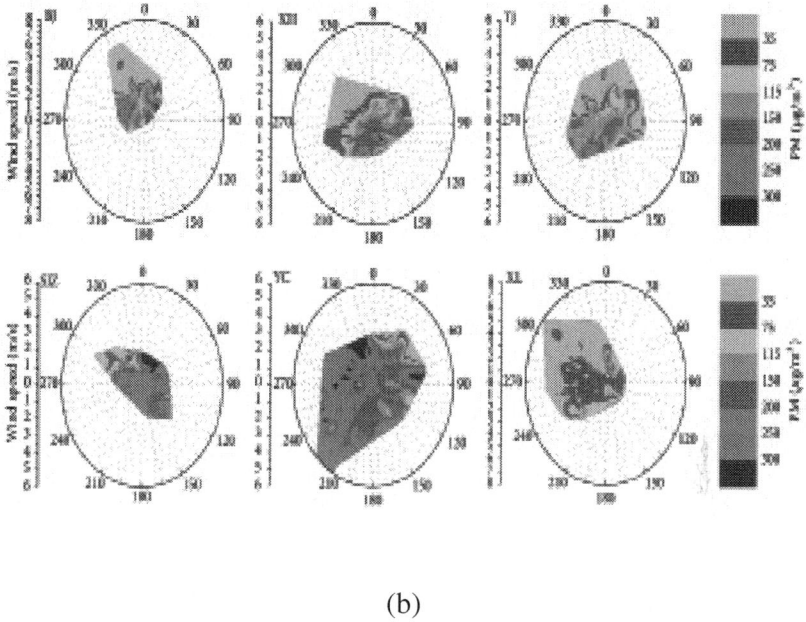

(b)

Figure 8: Maps of (a) wind roses and (b) relationships between hourly average PM$_{2.5}$concentration and wind speed.

Influence of Weather Systems, Horizontal Wind Field, and Structures of the Atmospheric Boundary Layer (ABL) on Consecutive and Regionally Severe Haze Episode

The heaviest regional haze pollution episode (January 9 to 14) is described in Figure 9. In contrast to other winter regional haze events reported by Zhao et al. [21]; the most significant characteristic of this episode was the "explosive growth" of PM$_{2.5}$ and PM$_{10}$ [6] on 12 January in Beijing. Synoptic maps and wind fields (Figure10), vertical profiles of wind (Figure 11), and temperature/humidity (Figure 12) were used to explore the formation of this heavy pollution from a meteorological perspective. On 10 January, the BTH region was located in the preceding part of low pressure and experienced southerly winds; the weak mixing and reduced dispersion under these conditions led to high levels of aerosol concentrations over the entire region. A north wind arrived north of Beijing due to moving low pressure at 14:00 LST on 11 January, and the particles in Beijing were quickly dispersed. However, the north wind only affected the air pollution in Beijing, and the south winds were still prevalent in the remainder of the region. Shortly afterward, a south wind occupied the entire region, and the $PM_{2.5}$ concentration in Beijing immediately rebounded due to the recirculation and transport of high-concentration pollutants from the south. Then, because of the combined effect of fast-moving weak low pressure in the north and steadily building high pressure in the south (Figure 10(c)), the winds in Beijing suddenly veered to the north and dispersed the pollutants at 02:00 LST on the 12th; the winds were soon southerly again at 14:00 LST. The pollutants were initially transported from Beijing but later returned. At the same time, the high-concentration pollutants in the south were transported into Beijing. Because the north wind only arrived in Beijing, the $PM_{2.5}$ concentrations in the remaining areas increased or stayed constant under the prevailing southerly winds. This recirculation and repeated accumulation and transport from the south over a short period were responsible for the explosive growth of the $PM_{2.5}$ and PM_{10}. A few hours later, east winds prevailed in the region, and the $PM_{2.5}$ concentrations at most of the sites began to decrease. However, Beijing was the convergence point of the weak southwest wind, southeast wind, and north wind (Figure 10(j)), mainly because of the urban heating effect under the weak pressure field. In addition, because the southeasterly winds transported warm humid air from the Bohai Sea, high humidity facilitated the hygroscopic growth of aerosols; as a result, severe low-visibility events ensued.

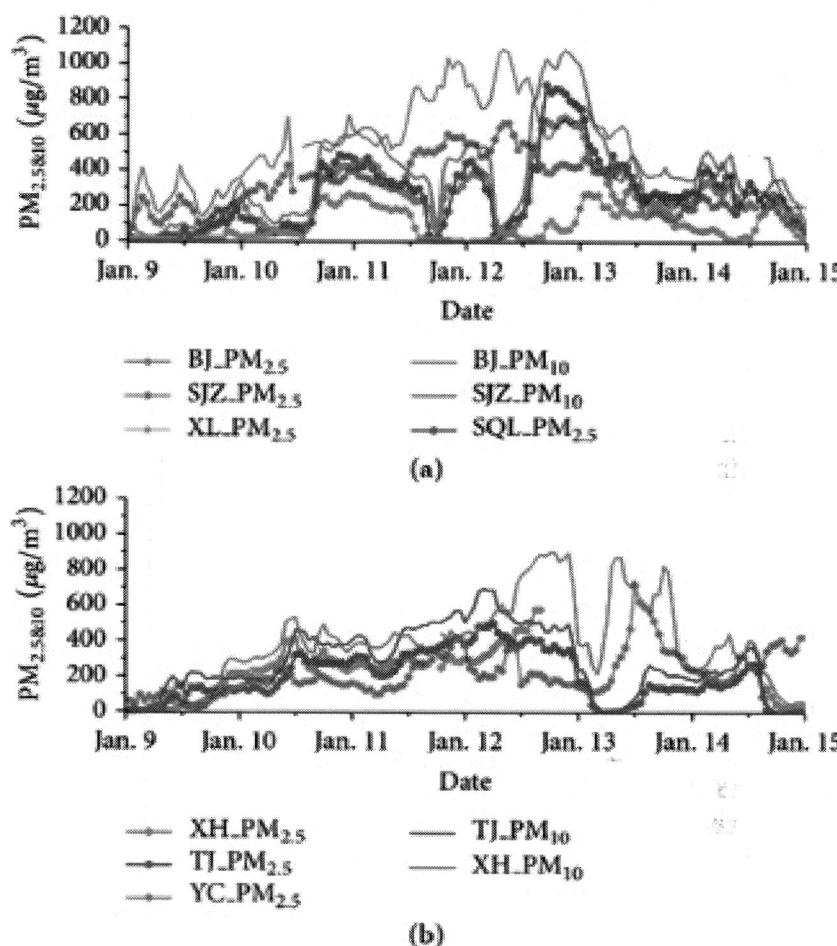

Figure 9: Variation of hourly average concentrations of PM$_{2.5}$ and PM$_{10}$ for different sites in the BTH region from January 9 to 14.

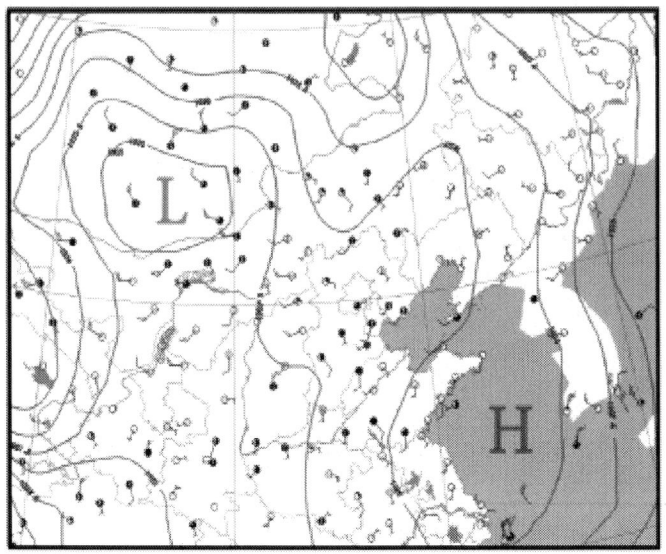

(a)

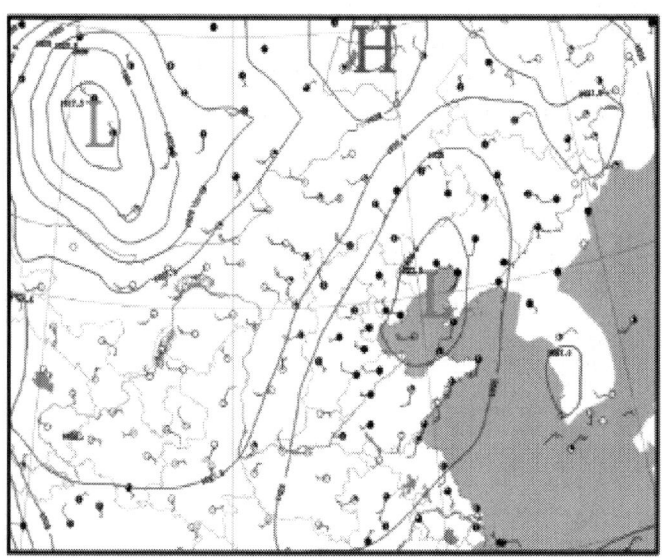

(b)

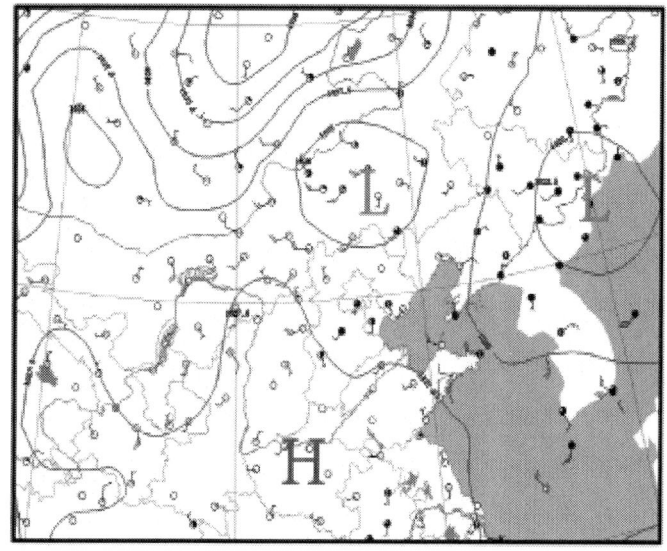

(c)

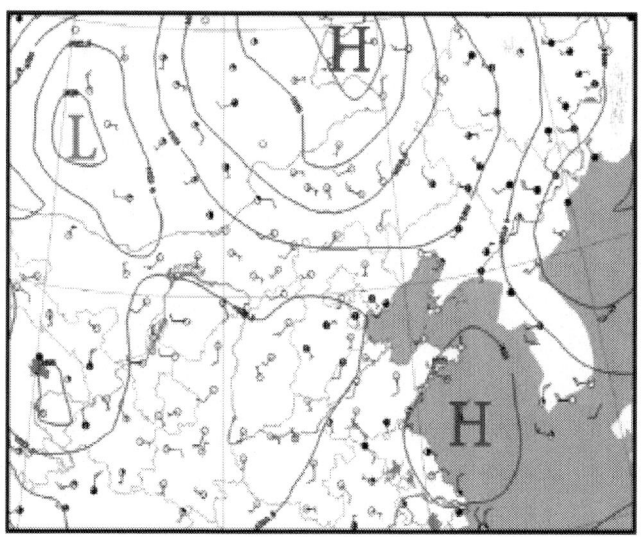

(d)

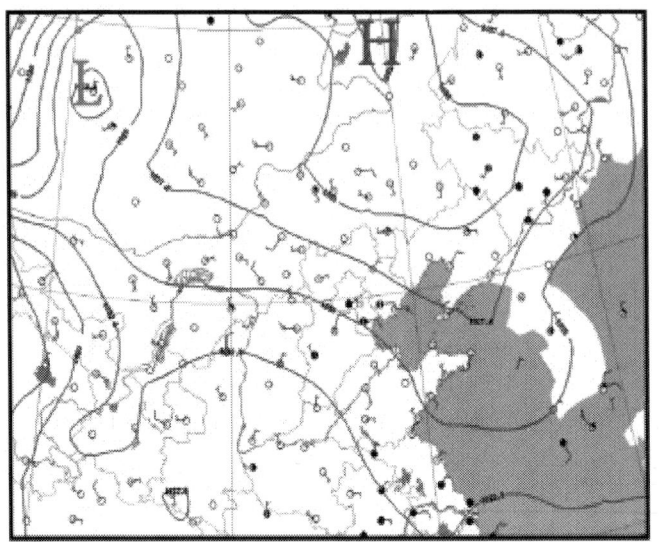

(e)

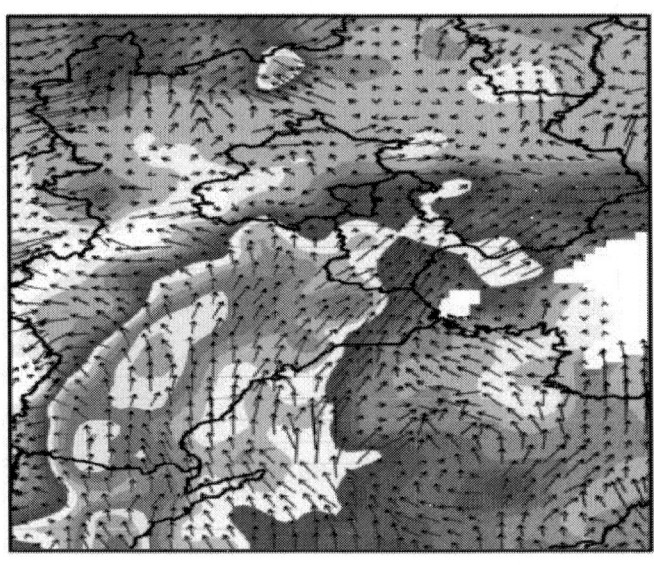

(f)

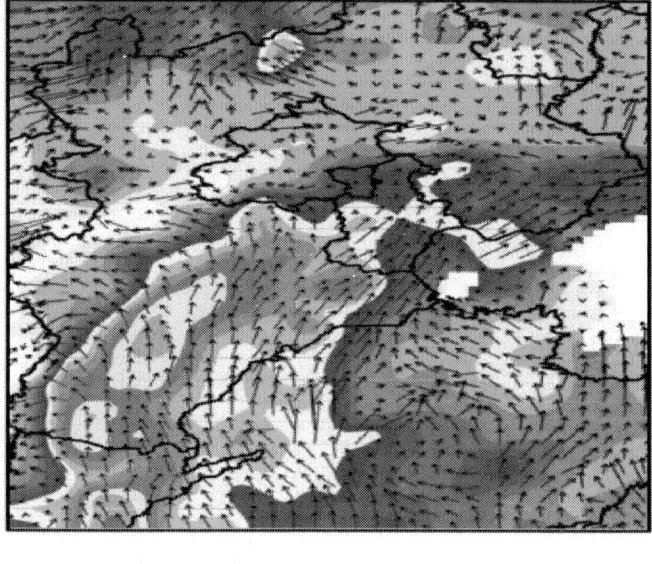

(g)

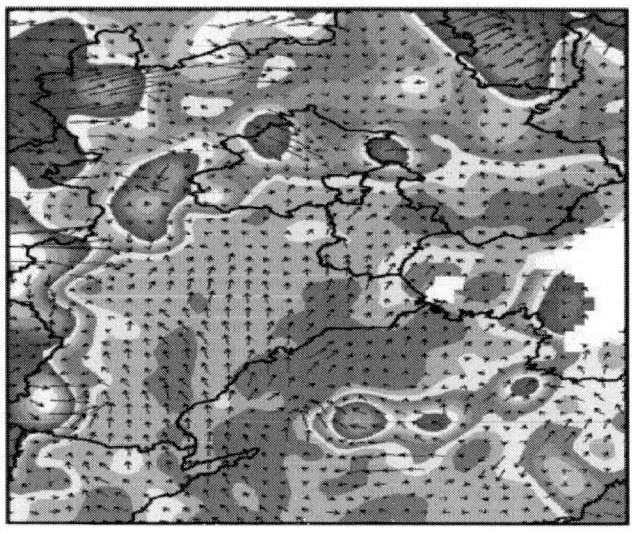

(h)

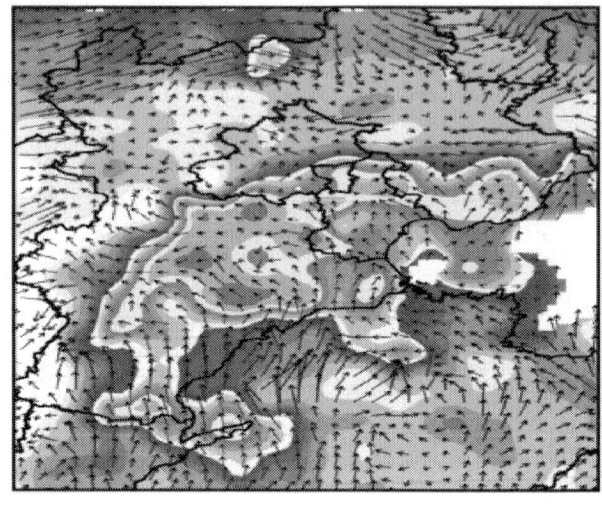

(i)

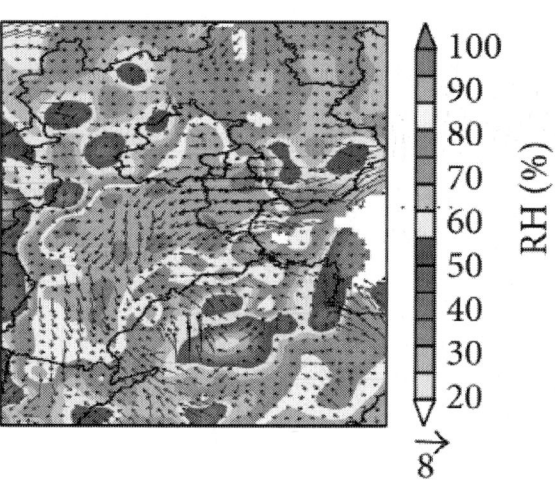

(j)

Figure 10: Synoptic maps at (a) 14:00 LST 10 January, (b) 14:00 LST 11 January, (c) 02:00 LST 12 January, (d) 14:00 LST 12 January, and (e) 02:00 LST

13 January; wind fields and humidity fields (shadings) at (f) 14:00 LST 10 January, (g) 17:00 LST 11 January, (h) 08:00 LST 12 January, (i) 14:00 LST 12 January, and (j) 02:00 LST 13 January.

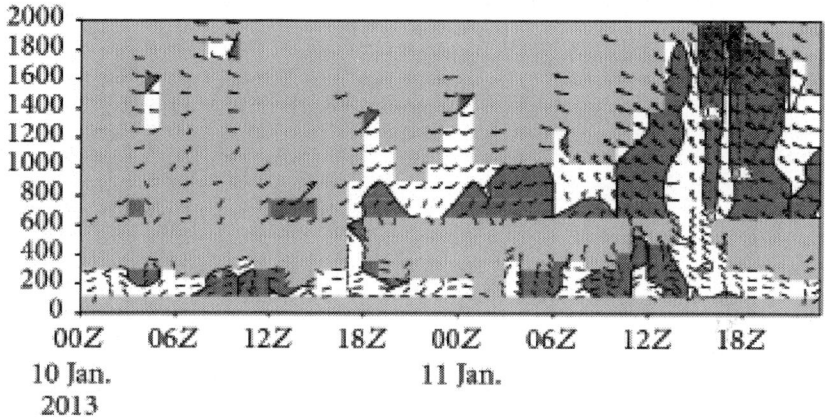

(a)

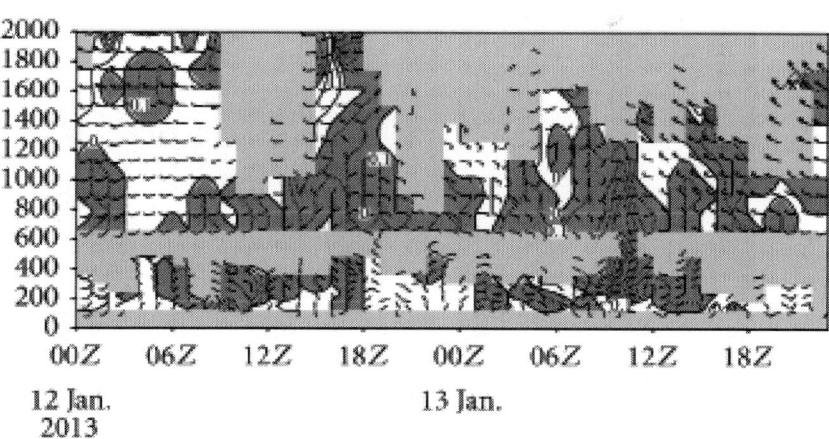

(b)

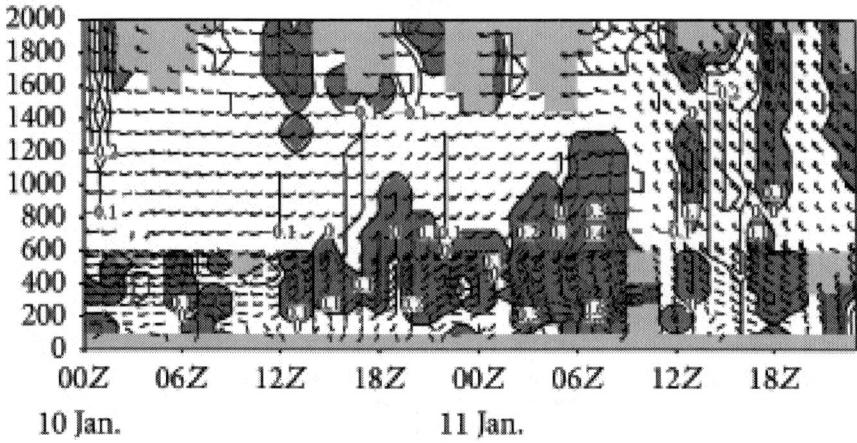

(c)

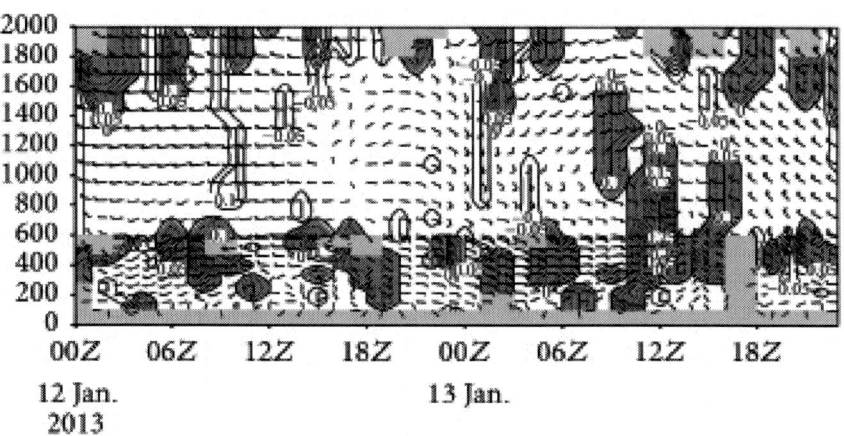

(d)

Figure 11: Wind profile (horizontal vector wind and vertical wind speed (unit: m/s)) in Beijing ((a) and (b)) and Tianjin ((c) and (d)) from January 10 to 13. The shaded map is vertical wind speed, white represents ascending motion with speed < 0 m/s, and gray represents descending motion with speed > 0 m/s.

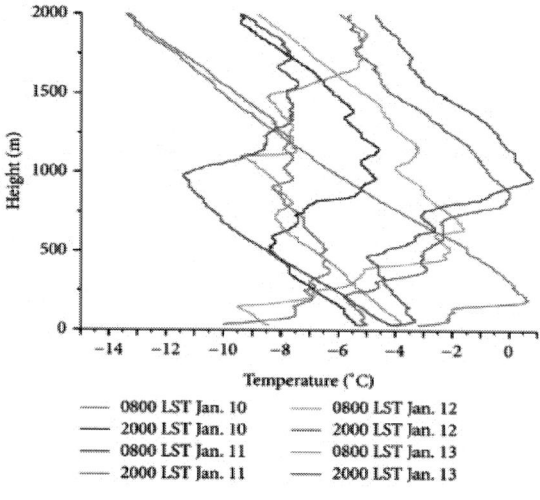

(a)

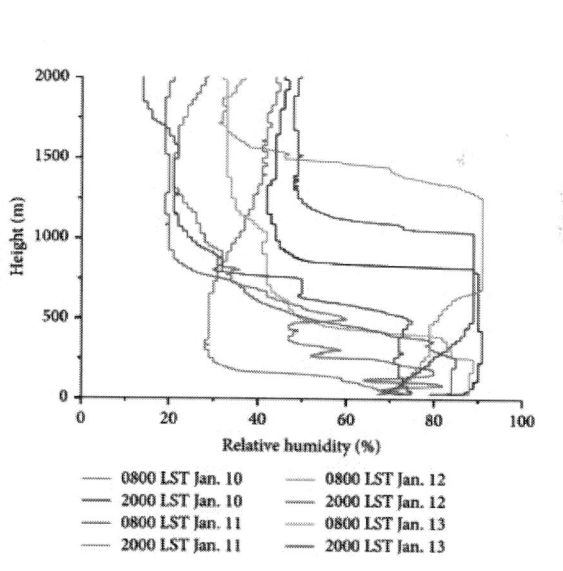

(b)

Figure 12: Profiles of temperature and humidity over different periods during 10th~13th January 2013, in Beijing.

The vertical wind profile (Figure 11) revealed very weak horizontal and vertical wind speeds from 14:00 LST on the 12th to 02:00 LST on the 13th in the Beijing urban region compared with those in the Tianjin suburban region. Moreover, two layers above and below 500 m were distinguishable; therefore, the mixed layer heights were approximately less than 500 m. Figure 12 shows the thickness of the inversion layer (600–800 m), the temperature range (5–7°C), and the high humidity forcing below 500 m on 12 January. Thus, recirculation and regional transport, along with the poorest diffusion conditions and favorable secondary transformation conditions under high emissions and the hygroscopic growth of aerosols, led to the explosive growth and the record high hourly average concentration of $PM_{2.5}$ in Beijing.

CONCLUSIONS

Haze pollution in winter over the Beijing-Tianjin-Hebei (BTH) region has become worse in recent decades. The most serious events occurred in January 2013, with a monthly regional average $PM_{2.5}$ value of ~145 μg/m³ and a visibility of ~8.3 km, which were most likely due to unfavorable meteorology, rather than an abrupt increase in emissions. The interannual variation in the mean visibility during haze days showed a significant decrease, which indicated that the primary type of haze has changed from slight haze (5–10 km) to severe or serious haze (<5 km) since the beginning of the 1990s due to unfavorable diffusion conditions (weak surface winds and high humidity) combined with high primary-pollutant emissions. The characteristics of the surface meteorological variables during moderate and severe haze pollution episodes in the BTH region are as follows: weaker pressure, higher temperature, particularly in the plains region, higher relative humidity, weak winds (1.7~2.1 m/s), fewer sunshine hours, and lower visibility. Orographic wind convergence zones resulted in the pollution accumulation in the piedmont plain and restrained the diffusion of pollutions; as a result, severe regional haze pollution developed. Recirculation and regional transport, along with the poorest diffusion conditions and favorable secondary transformation conditions under high emissions and the hygroscopic growth of aerosols, led to the explosive growth and the highest hourly average concentration of $PM_{2.5}$ in Beijing. Considering that decreasing wind speeds and weakened southerly winds resulted in more stable atmospheric conditions and weaker dispersion abilities, an effort should be made to control emissions and prevent haze events.

ACKNOWLEDGMENTS

This work was supported by National Natural Science Foundation of China (nos. 41230642 and 41175107) and the CAS Strategic Priority Research Program Grant (no. XDB05020203). The authors thank Hu Bo, Guiqian Tang, Yue Wang, Yang Yang, Na Chao, and Yangchun Yu for working for aerosol measures. They acknowledge National Meteorological Information Center, NOAA's National Climatic Data Center (NCDC), National Climate Center, and MODIS team for the data used in their work.

REFERENCES

1. M. Shao, X. Tang, Y. Zhang, and W. Li, "City clusters in China: air and surface water pollution,"Frontiers in Ecology and the Environment, vol. 4, no. 7, pp. 353–361, 2006.

2. G. Fu, W. Xu, R. Rong, J. Li, and C. Zhao, "The distribution and trends of fog and haze in the North China Plain over the past 30 years," Atmospheric Chemistry and Physics Discussions, vol. 14, no. 11, pp. 16123–16149, 2014.

3. National Bureau of Statistics of China, China Statistical Yearbook 2012, China Statistics Press, 2012.

4. J. Zhang, Z. Ouyang, H. Miao, and X. Wang, "Ambient air quality trends and driving factor analysis in Beijing, 1983–2007," Journal of Environmental Sciences, vol. 23, no. 12, pp. 2019–2028, 2011. · ·

5. P. Zhao, X. Zhang, X. Xu, and X. Zhao, "Long-term visibility trends and characteristics in the region of Beijing, Tianjin, and Hebei, China," Atmospheric Research, vol. 101, no. 3, pp. 711–718, 2011. · ·

6. Y. Wang, L. Yao, L. Wang et al., "Mechanism for the formation of the January 2013 heavy haze pollution episode over central and eastern China," Science China Earth Sciences, vol. 57, no. 1, pp. 14–25, 2014. · ·

7. D. Ji, Y. Wang, L. Wang et al., "Analysis of heavy pollution episodes in selected cities of northern China," Atmospheric Environment, vol. 50, pp. 338–348, 2012. ·

8. X. Zhuang, Y. Wang, H. He et al., "Haze insights and mitigation in China: an overview," Journal of Environmental Sciences, vol. 26, no. 1, pp. 2–12, 2014. · ·

9. X. Zhao, X. Zhang, X. Xu, J. Xu, W. Meng, and W. Pu, "Seasonal and diurnal variations of ambient PM2.5 concentration in urban and rural environments in Beijing," Atmospheric Environment, vol. 43, no. 18, pp. 2893–2900, 2009. · ·

10. Z. Li, C. Li, H. Chen et al., "East Asian Studies of Tropospheric Aerosols and their Impact on Regional Climate (EAST-AIRC): an overview," Journal of Geophysical Research D: Atmospheres, vol. 116, no. 4, Article ID D00K34, 2011. · ·

11. P. Li, J. Xin, Y. Wang, et al., "The acute effects of fine particles on respiratory mortality and morbidity in Beijing, 2004–2009," Environmental Science and Pollution Research, vol. 20, no. 9, pp. 6433–6444, 2013. · ·

12. B. Zhao, P. Wang, J. Z. Ma, S. Zhu, A. Pozzer, and W. Li, "A high-resolution emission inventory of primary pollutants for the Huabei region, China," Atmospheric Chemistry and Physics, vol. 12, no. 1, pp. 481–501, 2012. · ·

13. I. C. Ziomas, D. Melas, C. S. Zerefos, A. F. Bais, and A. G. Paliatsos, "Forecasting peak pollutant levels from meteorological variables," Atmospheric Environment, vol. 29, no. 24, pp. 3703–3711, 1995. · ·

14. J. H. Seinfeld and S. N. Pandis, Atmospheric Chemistry and Physics: From Air Pollution to Climate Change, John Wiley & Sons, New York, NY, USA, 1998.

15. M. Chin, "Atmospheric science: dirtier air from a weaker monsoon," Nature Geoscience, vol. 5, no. 7, pp. 449–450, 2012. · ·

16. J. Zhu, H. Liao, and J. Li, "Increases in aerosol concentrations over eastern China due to the decadal-scale weakening of the East Asian summer monsoon," Geophysical Research Letters, vol. 39, no. 9, Article ID L09809, 2012. · ·

17. Z. H. Chen, S. Y. Cheng, J. B. Li, X. R. Guo, W. H. Wang, and D. S. Chen, "Relationship between atmospheric pollution processes and synoptic pressure patterns in northern China," Atmospheric Environment, vol. 42, no. 24, pp. 6078–6087, 2008. ··

18. Y. Sun, T. Song, G. Tang, and Y. Wang, "The vertical distribution of PM2.5 and boundary-layer structure during summer haze in Beijing," Atmospheric Environment, vol. 74, pp. 413–421, 2013. ··

19. D. Ji, L. Li, Y. Wang et al., "The heaviest particulate air-pollution episodes occurred in northern China in January, 2013: insights gained from observation," Atmospheric Environment, vol. 92, pp. 546–556, 2014. ··

20. D. G. Streets, J. S. Fu, C. J. Jang et al., "Air quality during the 2008 Beijing Olympic Games," Atmospheric Environment, vol. 41, no. 3, pp. 480–492, 2007. ··

21. X. Zhao, P. Zhao, J. Xu et al., "Analysis of a winter regional haze event and its formation mechanism in the North China Plain," Atmospheric Chemistry and Physics, vol. 13, no. 11, pp. 5685–5696, 2013.

22. M. Tao, L. Chen, L. Su, and J. Tao, "Satellite observation of regional haze pollution over the North China Plain," Journal of Geophysical Research D: Atmospheres, vol. 117, no. 12, Article ID D12203, 2012. ··

23. J. Y. Xin, Y. S. Wang, G. Q. Tang et al., "Variability and reduction of atmospheric pollutants in Beijing and its surrounding area during the Beijing 2008 Olympic Games," Chinese Science Bulletin, vol. 55, no. 18, pp. 1937–1944, 2010. ··

24. H. Wang, S.-C. Tan, Y. Wang et al., "A multisource observation study of the severe prolonged regional haze episode over eastern China in January 2013," Atmospheric Environment, vol. 89, pp. 807–815, 2014. ··

25. H. He, Y. Wang, Q. Ma et al., "Mineral dust and NOx promote the conversion of SO_2 to sulfate in heavy pollution days," Scientific Reports, vol. 4, article 4172, 2014. ··

26. Y. L. Sun, Q. Jiang, Z. F. Wang et al., "Investigation of the sources and evolution processes of severe haze pollution in Beijing in January 2013," Journal of Geophysical Research-Atmospheres, vol. 119, no. 7, pp. 4380–4398, 2013.

27.   G. J. Zheng, F. K. Duan, Y. L. Ma, et al., "Exploring the severe winter haze in Beijing," Atmospheric Chemistry and Physics, vol. 14, no. 12, pp. 17907–17942, 2014.

28.   R. H. Zhang, Q. Li, and R. N. Zhang, "Meteorological conditions for the persistent severe fog and haze event over eastern China in January 2013," Science China Earth Sciences, vol. 57, no. 1, pp. 26–35, 2014. ··

29.   P. Zhao, X. Zhang, and X. Xu, "Comparison between two methods of distinguishing haze days with daily mean and 14 o›clock meteorological data," Acta Scientiae Circumstantiae, vol. 31, no. 4, pp. 704–708, 2011.

30.   Y. J. Kaufman, D. Tanré, L. A. Remer, E. F. Vermote, A. Chu, and B. N. Holben, "Operational remote sensing of tropospheric aerosol over land from EOS moderate resolution imaging spectroradiometer,"Journal of Geophysical Research D: Atmospheres, vol. 102, no. 14, pp. 17051–17067, 1997. ··

31.   D. Tanré, Y. J. Kaufman, M. Herman, and S. Mattoo, "Remote sensing of aerosol properties over oceans using the MODIS/ EOS spectral radiances," Journal of Geophysical Research D: Atmospheres, vol. 102, no. 14, pp. 16971–16988, 1997.

32.   L. A. Remer, Y. J. Kaufman, D. Tanré et al., "The MODIS aerosol algorithm, products, and validation,"Journal of the Atmospheric Sciences, vol. 62, no. 4, pp. 947–973, 2005. ··

33.   L. A. Remer, D. Tanré, and Y. J. Kaufman, "Algorithm for remote sensing of tropospheric aerosol form MODIS: collection 5," 2006, http://modis-atmos.gsfc.nasa.gov/MOD04_L2/.

34.   R. C. Levy, L. A. Remer, S. Mattoo, E. F. Vermote, and Y. J. Kaufman, "Second-generation operational algorithm: Retrieval of aerosol properties over land from inversion of Moderate Resolution Imaging Spectroradiometer spectral reflectance," Journal of Geophysical Research D: Atmospheres, vol. 112, no. 13, Article ID D13211, 2007. ··

35.   N. C. Hsu, S.-C. Tsay, M. D. King, and J. R. Herman, "Aerosol properties over bright-reflecting source regions," IEEE Transactions on Geoscience and Remote Sensing, vol. 42, no. 3, pp. 557–569, 2004. ··

36. H. Guo, M. Xu, and Q. Hu, "Changes in near-surface wind speed in China: 1969–2005," International Journal of Climatology, vol. 31, no. 3, pp. 349–358, 2011. · ·

37. Y. H. Yang, N. Zhao, X. H. Hao, and C. Q. Li, "Decreasing trend of sunshine hours and related driving forces in North China," Theoretical and Applied Climatology, vol. 97, no. 1-2, pp. 91–98, 2009. · ·

Chapter 6

Short- and Medium-Term Induced Ionization in the Earth Atmosphere by Galactic and Solar Cosmic RaysS

Alexander Mishev[1, 2]

[1]Institute for Nuclear Research and Nuclear Energy, Bulgarian Academy of Sciences, 1784 Sofia, Bulgaria

[2]Sodankyla Geophysical Observatory (Oulu Unit), University of Oulu, 90014 Oulu, Finland

ABSTRACT

The galactic cosmic rays are the main source of ionization in the troposphere of the Earth. Solar energetic particles of MeV energies cause an excess of ionization in the atmosphere, specifically over polar caps.

The ionization effect during the major ground level enhancement 69 on January 20, 2005 is studied at various time scales. The estimation of ion rate is based on a recent numerical model for cosmic-ray-induced ionization. The ionization effect in the Earth atmosphere is obtained on the basis of solar proton energy spectra, reconstructed from GOES 11 measurements and subsequent full Monte Carlo simulation of cosmic-ray-induced atmospheric cascade. The evolution of atmospheric cascade is performed with CORSIKA 6.990 code using FLUKA 2011 and QGSJET II hadron interaction models. The atmospheric ion rate is explicitly obtained for various latitudes, namely, 40°N, 60°N and 80°N. The time evolution of obtained ion rates is presented. The short- and medium-term ionization effect is compared with the average effect due to galactic cosmic rays. It is demonstrated that ionization effect is significant only in subpolar and polar atmosphere during the major ground level enhancement of January 20, 2005. It is negative in troposphere at midlatitude, because of the accompanying Forbush effect.

INTRODUCTION

Cosmic rays are high, ultrahigh, and extremely high energy particles of extraterrestrial origin, mostly protons. Cosmic rays (CRs) constantly impinge the Earth's atmosphere. While the low-energy particles are absorbed in the atmosphere, those with energies greater than 1 GeV/nucleon generate new particles through interactions with atomic nuclei. They are an important source of ionization in the Earth atmosphere [1]. The ionization in the stratosphere and troposphere is governed by galactic cosmic rays [2]. They initiate a complicated nuclear-electromagnetic-muon cascade resulting in an ionization of the ambient air. In such a cascade a small fraction of the initial primary particle energy reaches the ground as high energy secondary particles. Most of the primary energy is released in the atmosphere by ionization and excitation of the air molecules, resulting in an ionization of the ambient air. The maximum in secondary particle energy release is observed at altitudes of 15–26 km depending on latitude and solar activity level. This is the Pfotzer maximum.

The galactic cosmic ray (GCR) is affected by solar activity. They follow the 11-year solar cycle and respond to long and short time scale solar-wind variations. They are modulated with the opposite phase, that is, the higher solar activity, the lower the intensity of galactic CRs is. Solar energetic particles (SEPs) are accelerated during explosive energy release on the Sun and by acceleration processes in the interplanetary space. They enter the atmosphere sporadically, with a greater probability during periods of increased solar activity. In addition the heliosphere transient phenomena lead to a strong, relatively short suppression of GCR intensity in the vicinity of Earth, followed by a slower recovery on the time scale of several days known as Forbush decrease [3]. These events are generally interpreted as a result of the influence of coronal mass ejections (CMEs) and/or high-speed streams of the solar wind from the coronal holes on the background CRs.

The abundances of CR are nearly independent of the energy. For lower energies below 1 GeV/nucleon, the relative abundance of heavier nuclei increases, particularly around solar maximum, because they are less modulated than protons. In addition for a given energy, protons produce an atmospheric cascade that develops deeper in the atmosphere than cascades from heavier nuclei.

The investigation of ionization processes in atmosphere is important for better understanding of various processes and space weather mechanisms [1]. As example galactic cosmic rays influence via ionization the electrical parameters of planetary atmospheres. They are also related to atmospheric chemistry, as example the ozone depletion in the stratosphere. In addition to CR particles, solar electromagnetic X and UV radiations can affect the ionosphere and atmosphere, specifically the upper atmosphere. The effect varies geographically following the insolation pattern. It is beyond the topic of this study.

In the work presented here the ionization rate in the atmosphere during the major solar energetic particle event on January 20, 2005 is estimated. The obtained ion rates are compared with the average GCR ion production. This event is considered for study, because it is among the largest solar energetic particles events. In addition, the event occurred during the winter period; therefore the contribution of solar UV radiation could be neglected at polar latitudes for further atmospheric studies.

MODELING OF CR INDUCED ATMOSPHERIC IONIZATION

The estimation of cosmic-ray-induced ionization, as was recently demonstrated, could be performed on the basis of a full Monte Carlo simulation of the atmospheric cascade [4]. At present, with the development of numerical methods, evolution of the knowledge of high energy interactions and nuclear processes, an essential progress in models for cosmic-ray-induced ionization in the Earth atmosphere is carried out [5–8]. As was recently demonstrated the models agree with 10–20%, the difference is mainly due to the various hadron generators and atmospheric models [2].

The full Monte Carlo simulation of the atmospheric cascade permits to follow the longitudinal cascade evolution in the atmosphere and obtains the energy deposit by different atmospheric cascade components from ground level till the upper atmosphere. These full target models apply the formalism of ionization yield function Y

$$Y(x, E) = \Delta E(x, E) \frac{1}{\Delta x} \cdot \frac{1}{E_{ion}} \cdot \Omega,$$

(1)

where ΔE is the deposited energy in layer x in the atmosphere and Ω is a geometry factor, integration over solid angle, $E_{ion} = 35eV$, which is the energy necessary for production of one ion pair [9, 10]. The ionization yield function represents the number of ion pairs produced at given altitude x in the atmosphere by given primary cosmic ray nuclei with kinetic energy E on the top of the atmosphere. In fact the ionization yield function represents the ionization capacity in air by given primary CR particle.

The estimation of atmospheric ion rate production is obtained by convolution of Y(x, E) and SEP spectrum as shown in

$$q(h, \lambda_m) = \int_{E_0}^{\infty} D(E, \lambda_m) Y(h, E) \rho(h) dE,$$

(2)

Where $D(E, \lambda_m)$ is the differential primary cosmic ray spectrum at given geomagnetic latitude λ_m for a given component of primary cosmic ray, y is the yield function (1), and p(h) is the atmospheric density $(g \cdot cm^{-3})$. We express x, during the simulations in $g \cdot cm^{-2}$, which is a residual atmospheric depth, that is, the amount of matter (air) overburden above a given altitude in the atmosphere. This is naturally related to the development of the cascade. Subsequently the mass overburden is transformed as altitude above sea level (a.s.l.) in [km].

In this study the simulation of the development of atmospheric cascade is carried out with CORSIKA 6.990 code [11] with corresponding hadron interaction generators FLUKA 2011 [12] and Quark Gluon String with JETs QGSJET II [13]. COsmic Ray SImulations for KASKADE (CORSIKA) code is the most widely used atmospheric cascade simulation tool. The code simulates the interactions and decays of various nuclei, hadrons, muons, electrons, and photons in the atmosphere. The particles are tracked through the atmosphere until they undergo reactions with an air nucleus or in the case of unstable secondary particles, they decay. The result of the simulations is detailed information about the type, energy, momenta, location, and arrival time of the produced secondary particles at given selected altitude a.s.l. The primary particles that can be considered are protons, light, middle, and heavy nuclei up to iron.

IONIZATION EFFECT IN THE ATMOSPHERE DUE TO GALACTIC AND SEP

The described above formalism is applicable for the whole atmosphere where CR could affect from ground level to altitude of 100 km a.s.l. After obtaining the energy deposition from all secondary cosmic ray particles, respectively, ionization yield function y (1) on the basis of (2), the ion rate as ion pairs per second in cm^3 is obtained. The ion rates are presented as ion pairs per second, per cm^3, and per atmosphere.

Ion Rates at Solar Minimum and Maximum at Various Atmospheric Depths

The cosmic-ray-induced ionization by GCR at solar minimum and maximum is obtained for ground level to the upper atmosphere. In Figure 1 the obtained ion rates are presented for various atmospheric depths, namely, $100\,g\,cm^{-2}$ (16 km a.s.l.), $300\,g \cdot cm^{-2}$ (10 km a.s.l.), $500\,g \cdot cm^{-2}$ (5 km a.s.l.), and $700\,g \cdot cm^{-2}$ (3 km a.s.l.) as a function of the rigidity cut-off.

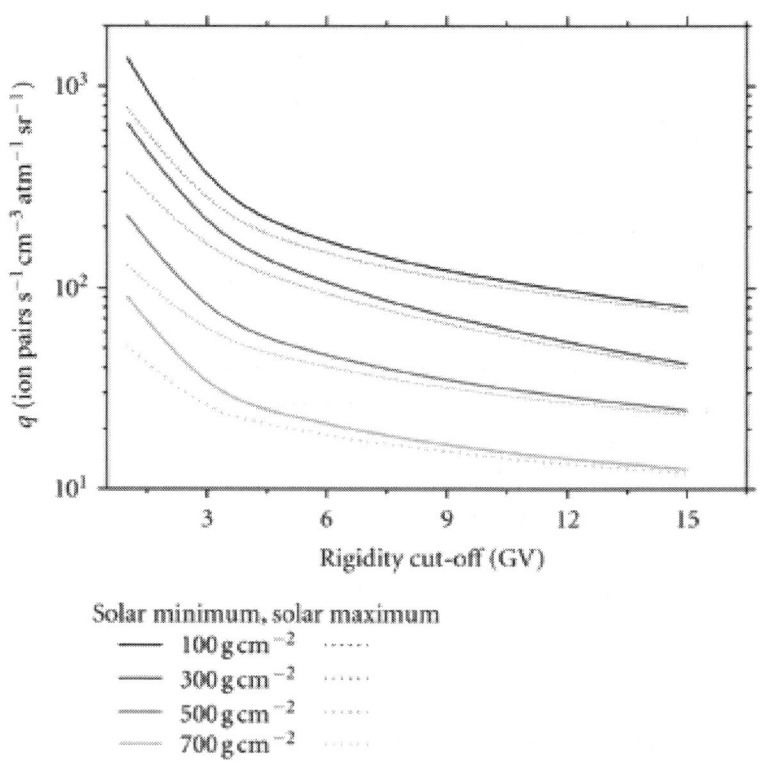

Figure 1: Cosmic-ray-induced ionization by GCR at solar minimum and maximum at various atmospheric depths.

It is demonstrated that solar activity affects ion rate production significantly at high latitudes, while at the Equator (15 GV cut-off), this effect is not so strong. Such computation of GCR-induced ionization permits to estimate the ion rate production at various time scales. Therefore the average ion rate due to GCR could be estimated for specific conditions (solar minimum and maximum). Hence, the obtained ion rates due to GCR permit to compare the SEP ionization effect with average ionization CR effect in a realistic manner.

Ionization Effect on January 20, 2005, During the Major GLE 69

In addition to continuous ionization in the Earth's atmosphere caused by galactic cosmic ray a sporadic ionization occurred during solar energetic particle events, potentially affecting the Earth's environment [14]. In general such events are low energy and are not able to initiate atmospheric cascades. Their ionization effect is limited to the upper polar atmosphere. Hence the studies of the effects caused by solar energetic particles are usually limited to the upper atmosphere above 30 km. However, it was recently demonstrated [15–18] that during some major ground level enhancements (GLEs), which are characterized by very high energy of solar particles, the ionization effect is important specifically over polar atmosphere.

On January 20, 2005 all energy channels of GOES satellite simultaneously register enhancement of proton flux. The (SEP) onset was registered at 6:50 UT, 3 minutes before maximum of X-ray flare. During the first hour the proton spectrum parameters change dramatically [19]. The changes of the spectra could be connected with some particle acceleration mechanism preceding the CME launch. The next 10 hours of acceleration produce spectra with very stable parameters [20], most likely formed at CME driven shock site.

A well-studied event [19] such as ground level enhancement 69 on January 20, 2005 was considered for analysis. Since this event is among the largest (the second largest in observation history), the ionization effect is expected to be maximal. The event on January 20, 2005 is characterized by an anisotropic component with a very hard spectrum at the onset of the event, followed by a long isotropic emission with a softer spectrum [21]. The spectral index for SEP during GLE is typically between 4 and 6. In this work the solar proton spectrum is obtained on the basis of GOES 11 satellite measurements (high energy channels) and additional data [22, 23].

The spectrum of solar protons is expressed in two different moments: at 08:00 UT a high energy part with a slope of 2.32 and at 23:00 UT low energy part with a slope of 3.43. During the simulation, a realistic winter profile of the Earth atmosphere is considered [24, 25]. It is demonstrated that the ionization effect on event onset at 08:00 UT (Figure 2(a)) is greater than the one produced by delayed component at 23:00 UT (Figure2 (b)). Because the event on January 20, 2005 occurred during the recovery phase of the Forbush decrease and in the following days an additional suppression of the cosmic ray intensity was observed leading to a complicated time profile of cosmic ray flux, the net ionization effect is calculated as a superposition of ion rate from solar energetic particles and from reduced galactic cosmic rays.

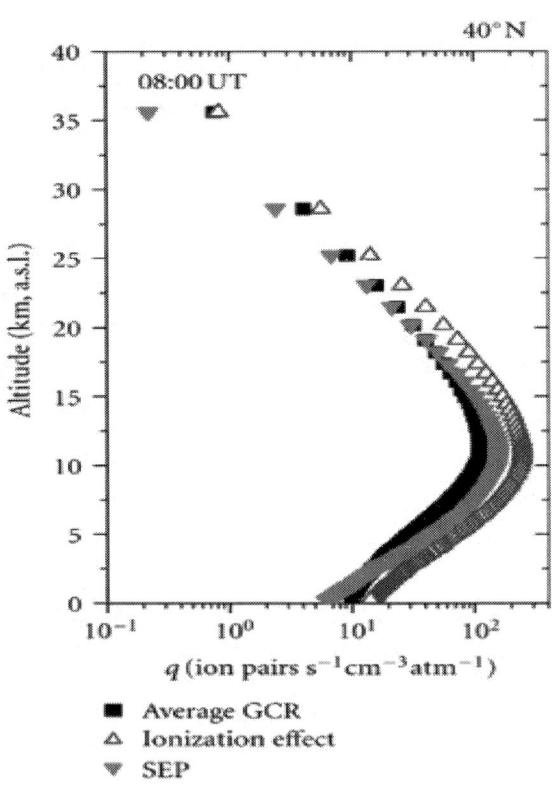

(a)

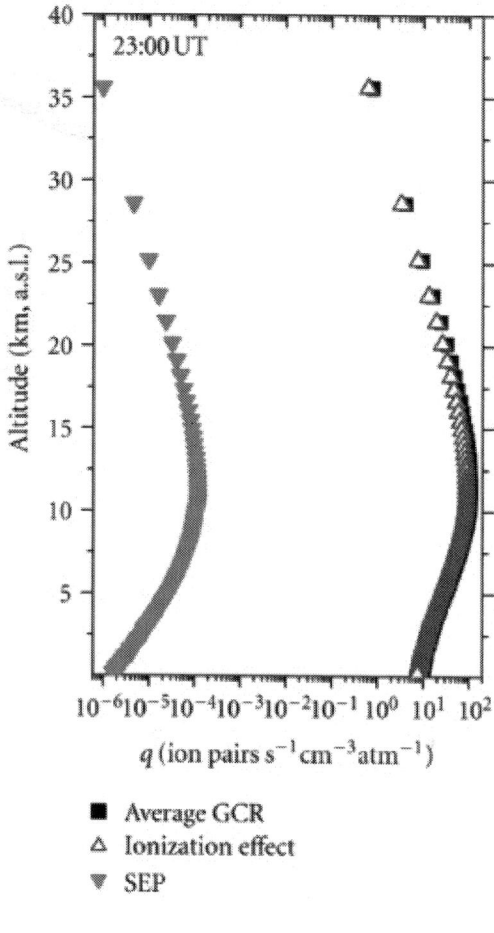

(b)

Figure 2: Ionization effect at 40°N due to GCR and SCR during GLE 69 on January 20, 2005.

In the case of 40°N latitude the effect at 08:00 UT is comparable to the average of GCR (Figures 1 and 2(a)). However, the ion rates quickly decrease with altitude (below 5 km a.s.l.). The ionization effect due to low energy component of the SEP spectrum, namely, at 23:00 UT is negligible (Figure 2(b)). In this case the ion rates are due mostly from reduced GCR.

The situation is quite different for latitude of 60°N (Figure 3). The ion rate due to SEP at 08:00 UT is significant. The ion rates from solar particles are greater than ion rates from GCR roughly an order of magnitude. The ionization is significant at altitudes above some 12 km a.s.l. and decreases in the troposphere. The ion rate at 23:00 UT due to a low energy component, as in a previous case, is negligible. In the case of 80°N latitude both components, hard at 08:00 UT, respectively, soft at 23:00 UT cause a significant excess on ion rates in the atmosphere The effect at 08:00 UT is due mainly to solar protons (Figure 4(a)). It is significant at altitudes above 10–12 km a.s.l. Because the hard spectrum is with a slope of 2.32, it causes significant ionization even in a low atmosphere. At 23:00 UT the effect due to solar particles is significant at altitudes above 12 km a.s.l. and decreases in the troposphere, because of the softer spectrum. The ionization effect in a low troposphere is due to a reduced GCR (Figure 4(b)). Computed ion rates permit to estimate the net ionization effect at various time scales and to compare it to the average effect due to GCR.

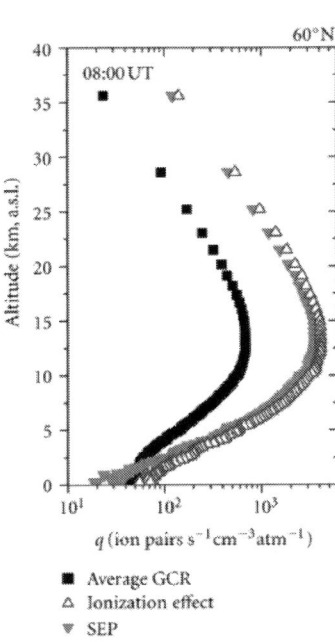

(a)

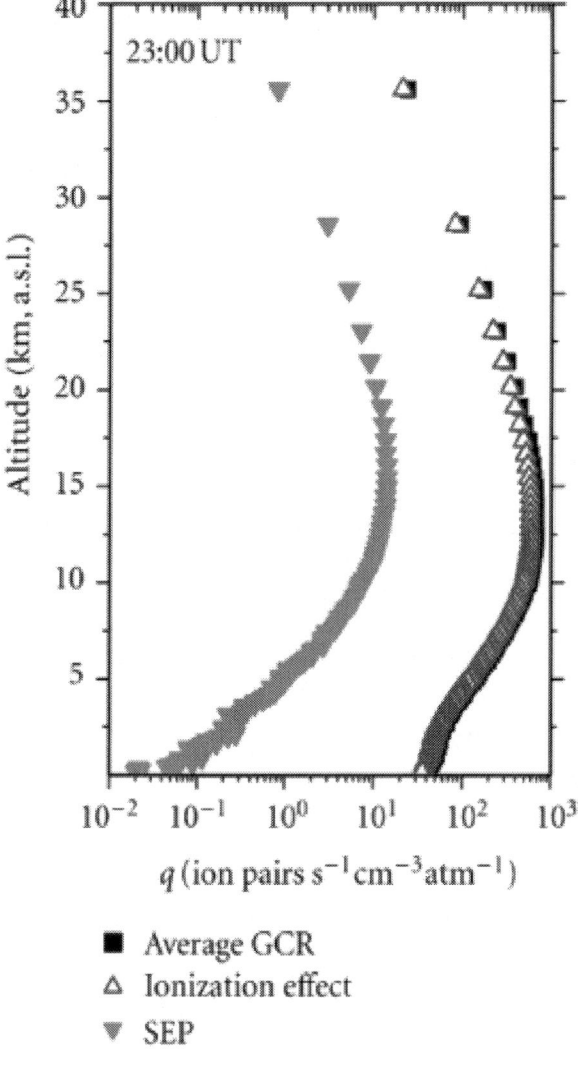

(b)

Figure 3: Ionization effect at 60°N due to GCR and SCR during GLE 69 on January 20, 2005.

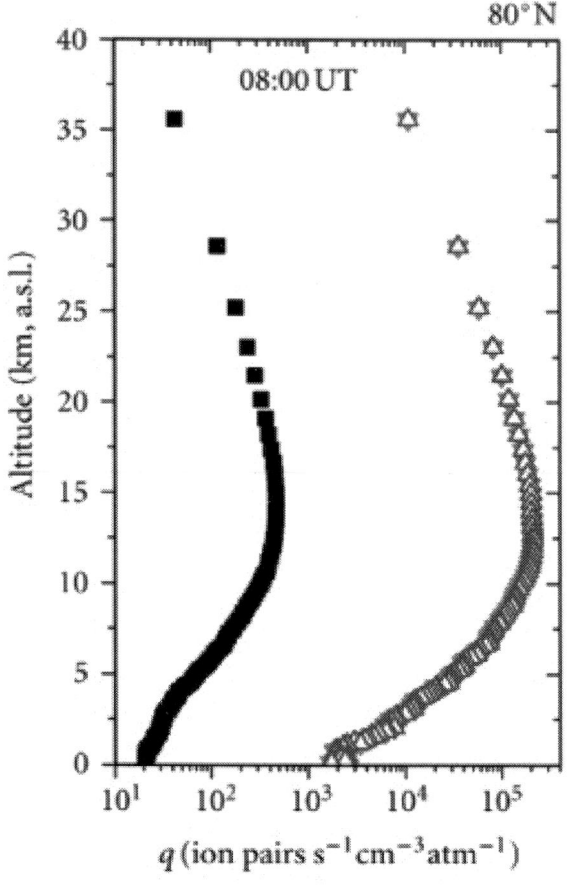

(a)

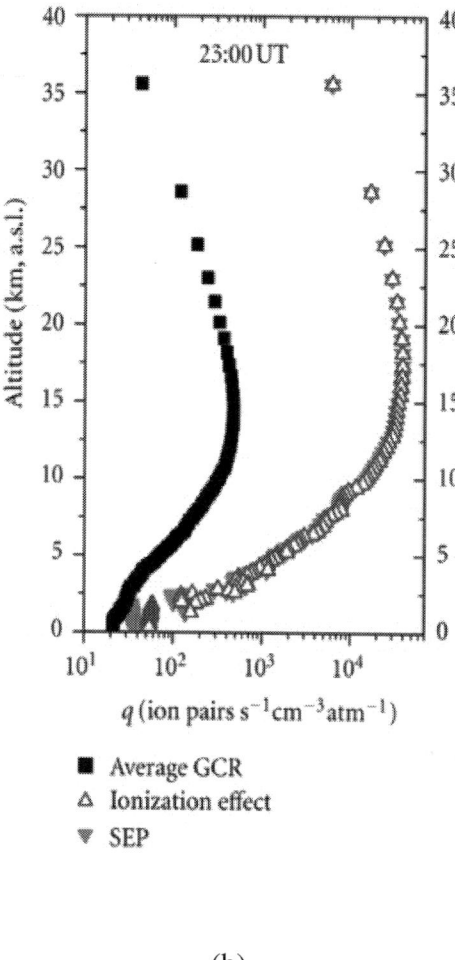

(b)

Figure 4: Ionization effect at 80°N due to GCR and SCR during GLE 69 on January 20, 2005.

Short and Medium Time Ionization Effect due to GLE 69 on January 20, 2005

The ionization effect is maximal during the first 24 h from the event onset [14, 16, 17]. The 24 h effect is computed as superposition of SEP and GCR ionization. Since the event occurred during the recovery phase of a strong Forbush decrease, a reduced GCR flux is considered. The 48 h effect is superposition of 24 h effect and reduced GCR ionization effect for the following 24 h. The weekly effect is computed taking into account the 24 h ionization effect from solar protons and complicated GCR flux, considering explicitly the transient phenomena during the week.

The ionization effect at various timescales is computed at various altitudes a.s.l., namely, at $100 \, g \, cm^{-2}$ (16 km a.s.l.), $300 \, g \cdot cm^{-2}$ (10 km a.s.l.), $500 \, g \cdot cm^{-2}$ (5 km a.s.l.), and $700 \, g \cdot cm^{-2}$ (3 km a.s.l.) as a function of the rigidity cut-off. This permits to estimate with good precision in which region of the atmosphere the ionization effect is significant, weak or negative. The Figures 5–7 demonstrate the net ionization effect as a function of the rigidity cut-off at various altitudes compared with average ionization effect from GCR. The average ionization effect from GCR is computed for period corresponding to the week before GLE 69 onset.

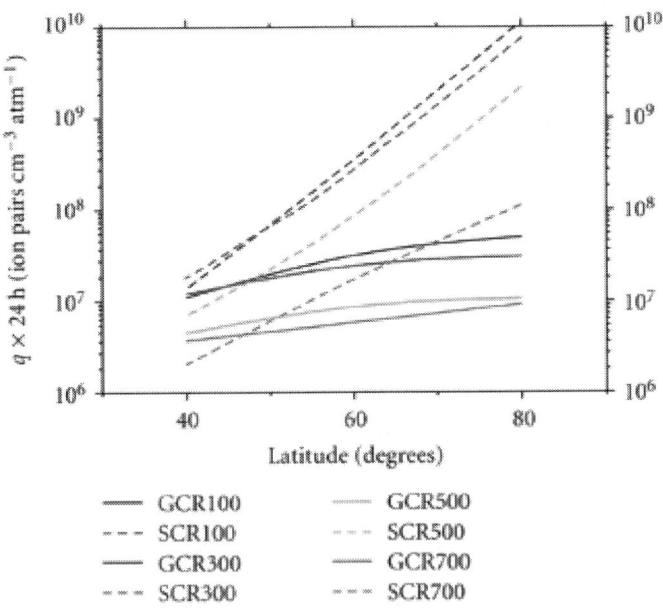

Figure 5: 24 h ionization effect due to solar protons during GLE69, compared with average GCR, as a function of latitude for various observation depths.

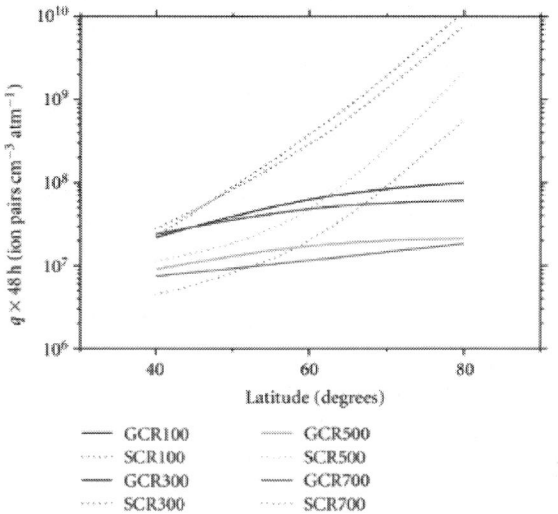

Figure 6: 48 h ionization effect due to solar protons during GLE69, compared with average GCR, as a function of latitude for various observation depths.

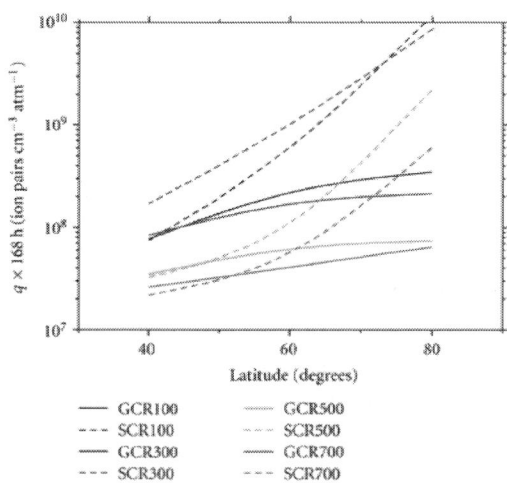

Figure 7: Weekly ionization effect due to solar protons during GLE69, compared with average GCR, as a function of latitude for various observation depths.

In general, as was shown in (Figure 2), the effect at 40°N is weak, with excess in the region of Pfotzer maximum. Therefore the 24 h is weak in a whole midlatitude atmosphere (Figure 5). We observe negative ionization effect compared to average effect from GCR at $700 \, g \cdot cm^{-2}$ (red lines on Figure 5). The effect from SEP is weak at $500 \, g \cdot cm^{-2}$ (green lines). Only at altitudes around some 10 km a.s.l. the ionization effect due to solar protons is greater than the average due to GCR at 40°N latitude. As was expected the effect increases for low rigidity cut-offs, that is, in subpolar and Polar Regions. We conclude that the 24 h ionization effect is important in the subpolar and polar atmosphere during major ground level enhancement of January 20, 2005, specifically at high altitudes (Figure 5—black and blue dashed lines).

The 48 h ion production at 40°N is comparable with the average due to GCR at altitudes above $500 \, g \cdot cm^{-2}$ (Figure 6). The 48 h ion production is below the average due to GCR at altitude of 3 km for 40°N. Because the significant ion production at 60°N and 80°N; the 48 h ionization effect is still strong at these regions, specifically at high altitudes. The 48 h effect results on 50% increase compared to the average due to GCR at 60°N. The effect is even stronger at 80°N. Therefore the 48 h ionization effect is important only in the subpolar and polar atmosphere during major ground level enhancement of January 20, 2005 at high altitudes.

The weekly effect at 40°N (Figure 7) is clearly negative at low altitude (green and red lines at Figure 7). It is comparable with average GCR (black lines at Figure 7) effect at high altitude. However, the significant ion production during the event onset results on important ionization effect in the region of the Pfotzer maximum at 40°N (blue lines at Figure 7). The weekly effect is weak at the troposphere even at subpolar latitudes. However at 80°N the weekly effect is still important.

An illustration of the above conclusions is shown in Figure 8. Figure 8 demonstrates the relativity to the average GCR ionization effect for 24 h, 48 h, and 168 h at 40°N in the low troposphere. It is shown that the 24 h ionization effect is positive in a tight region of the troposphere above some 5 km a.s.l. In this region the interplay between reduced GCR flux and SEP leads to slight increase of the ionization. As was expected the 48 h is negative, as well as the weekly ionization effect (red and blue lines). In the upper troposphere the 24 h ionization effect is positive compared to the average of GCR only is a small region (Figure 9). The ionization effect is negative in the stratosphere. The 48 h ionization effect is clearly negative as was mentioned above.

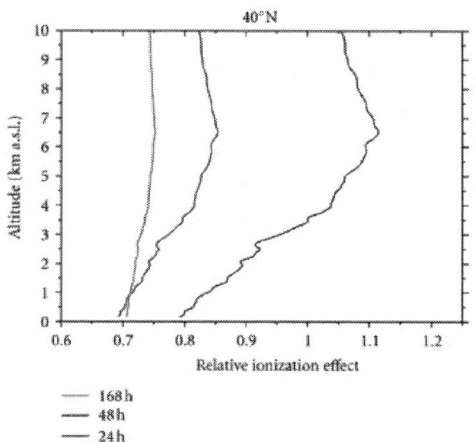

Figure 8: Relative to the average GCR ionization effect for 24 h, 48 h, and 168 h at 40°N in troposphere.

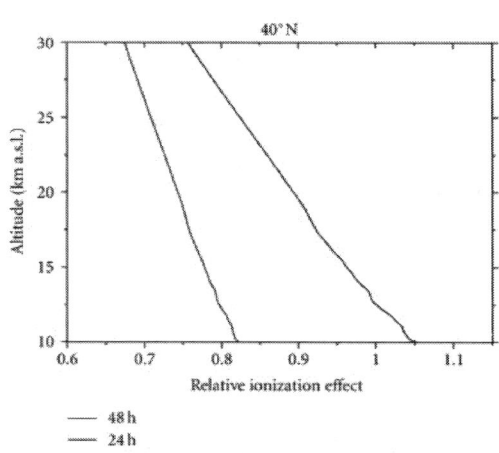

Figure 9: Relative to the average GCR ionization effect for 24 h and 48 h at 40°N in the upper troposphere and stratosphere.

Similarly it is demonstrated that 24 ionization effects at 60°N is not important in the lower troposphere (Figure10). The 48 h ionization effect is not important as well and is slightly negative in planetary boundary layer.

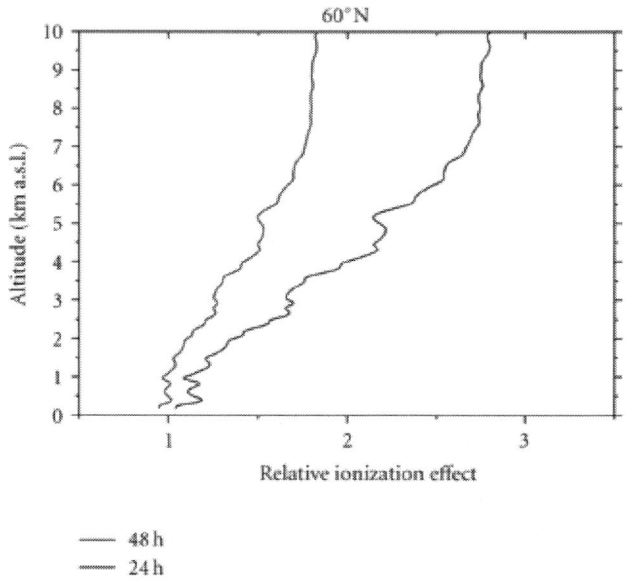

Figure 10: Relative to the average GCR ionization effect for 24 h and 48 h at 60°N in the troposphere.

Taking into account the time evolution of the obtained ion rates, the conclusion is the ionization effect is nearly negative for 40°N, especially in the troposphere and small for 60°N, because of the accompanying Forbush decrease during the event. The ionization effect is important only in the subpolar and polar atmosphere during major ground level enhancement of January 20, 2005 at high altitudes.

APPLICATIONS AND DISCUSSION

The application of recent Monte Carlo CR ionization models is related to the study of various processes in the atmosphere, specifically related to atmospheric chemistry and physics [26, 27]. The influence of GCR should not be neglected in investigations of the tropospheric and stratospheric chemistry and dynamics. The CR ionization models show that the effects of galactic cosmic rays on the atmosphere are statistically significant in large geographic regions and for a number of relevant atmospheric species.

It was recently demonstrated for Southern hemispheric troposphere that NO_x increases more than 20% in the polar region [26]; HO_x decreases of 3% in the midlatitude upper troposphere; HNO_3 increases by more than 10% between the South Pole and subtropics and O_3 increases by up to 3% throughout the troposphere to 20 km between the South Pole to 20°N. Respectively, for Northern hemispheric troposphere HNO_3 marginally significant increases in the midlatitude upper troposphere and O_3 significant decreases in the polar upper troposphere. Undoubtedly these effects are related with the planetary distribution of GCR.

In addition it was demonstrated that major solar events leading to GLEs could also impact the Earth atmosphere [28]. According to the current knowledge, the cosmic-ray-induced ionization of the atmosphere, due to solar and galactic cosmic rays produces additional chemical sources and influences ozone molecules in the atmosphere [29] via photochemistry [30]. Recently, was demonstrated, that the ionization of the polar atmosphere by SEPs is responsible for the impact on neutral constituents. In addition it was shown that even events with limited particle flux could change the abundance of the minor constituents in the mesosphere and the upper stratosphere [31].

The obtained results presented here are related to these studies, since the atmospheric ionization is related to the mentioned above phenomena. Recent studies at the upper atmosphere on short-medium time [32, 33] demonstrate that the January 2005 SPEs caused large enhancements in the northern polar mesospheric HO_x and NO_x constituents (both observed and modeled). Observations indicated large mesospheric increases in OH (up to 4 ppbv) and HO_2 (>0.5 ppbv) as a result of the SPEs during the time period 16–21 January in the 60–85N latitude. The WACCM3 simulations showed quantitatively similar, although somewhat larger enhancements in OH and HO_2. These large HO_x enhancements led to considerable ozone decreases of greater than 40% throughout most of the northern polar mesosphere during the SPE period. All these studies demonstrate that SEP impact on the atmosphere is more important than GLE particles in the upper atmosphere. However, for better understanding of the atmospheric chemistry and physics process is the investigation if their impact is important.

While the recent model studies of cosmic-ray-induced ionization during the major GLE 69 on January 20, 2005 focus on the onset of the event and the next following hours, when the effect is maximal, in the study presented here we focus on short- to mid-term ionization effect. Previous works did not concern the mid-term ionization effect in details for various altitudes in the atmosphere. In addition several of the previous studies are limited to a high atmosphere [32] or only on the contribution of heavy ions [34]. Therefore the results presented here extend the recent studies of such events and give basis for further improvement of model studies related to atmospheric physics phenomena.

CONCLUSIONS

The application of Monte Carlo methods for investigation of cosmic-ray-induced ionization is very important, because it is possible to consider explicitly the hadron component. It gives the possibility to estimate middle and low atmospheric effects related to cosmic-ray-induced ionization realistically. As was recently demonstrated their application in a specific conditions [15, 17] permits detailed study of the ionization effect, specifically at middle and low altitudes [16].

The effect of sporadic solar energetic particle events is limited on a global scale but most energetic events could be strong locally, specifically in subpolar and Polar Regions, affecting the physical-chemical properties of the upper atmosphere. Since the large SEP events leading to GLEs are different on spectra and composition, their detailed study is connected with detailed information about heliospheric and geospace conditions. In this connection extension of the existing models to the upper atmosphere is very important, as well as their comparison with analytical models.

The ionization effects in the upper atmosphere during GLE 69 on January 20, 2005 are well studied [32, 33]. In this study it is shown that ionization effect is significant at subpolar and polar atmosphere, with fast tropospheric decrease [16, 17]. Previous studies are focused on a maximal phase of ionization effect during the first hours from GLE onset. In this study the obtained ion rates are used to estimate the mid-term ionization effect. It is demonstrated that the ionization effect is nearly negative compared to GCR at middle latitudes for short- to mid-term periods. This is very important for further studies of the influence of GLE particles on atmospheric chemistry and physics.

It has a general agreement that cosmic-ray-induced ionization of the atmosphere, due to solar and galactic cosmic rays, produces additional chemical sources and influences ozone molecules in the atmosphere. As was mentioned above the ionization effect due to SEPs is important only in a subpolar and polar atmosphere. This is very important for recent composition changes studies. The January 20, 2005 SPEs caused large enhancements in the northern polar HO_x and NO_x constituents in mesosphere, which lead to ozone decrease in the order of 40%. However on the basis of the obtained ion rate it is demonstrated that the tropospheric and stratospheric ionization effects due to SEPs are important on short to medium time scales in polar atmosphere and are nearly negative compared to the average GCR on medium time scales at middle latitudes (because of the accompanying Forbush decrease). More detailed studies of the influence on minor components of the atmosphere deserve special interest.

In addition, it was recently demonstrated [34] that the contribution of light nuclei, specifically Helium could be important for total ionization even at middle latitudes during GLE 69. The extension of such models, specifically in the upper atmosphere on the basis of a full Monte Carlo simulation of the atmospheric cascade with combination with analytical procedures [35] will contribute to deep understanding of the impact of solar and galactic cosmic rays on the atmosphere of the Earth and their connection to the space weather [36–38] and atmospheric physics [39].

ACKNOWLEDGMENTS

The author warmly acknowledges the high energy Division of the Institute for Nuclear Research and Nuclear Energy of Bulgarian Academy of Sciences, namely, V. Guenchev, S. Piperov, E. Puncheva, and P. Konstantitov for the given computational time.

REFERENCES

1. G. A. Bazilevskaya, I. G. Usoskin, E. O. Flückiger et al., "Cosmic ray induced ion production in the atmosphere," Space Science Reviews, vol. 137, no. 1–4, pp. 149–173, 2008

2. I. G. Usoskin, L. Desorgher, P. Velinov et al., "Ionization of the earth's atmosphere by solar and galactic cosmic rays," Acta Geophysica, vol. 57, no. 1, pp. 88–101, 2009. · ·

3. S. E. Forbush, "Cosmic-ray intensity variations during two solar cycles," Journal of Geophysical Research, vol. 63, no. 4, pp. 651–669, 1958.

4. I. G. Usoskin, O. G. Gladysheva, and G. A. Kovaltsov, "Cosmic ray-induced ionization in the atmosphere: spatial and temporal changes," Journal of Atmospheric and Solar-Terrestrial Physics, vol. 66, no. 18, pp. 1791–1796, 2004. · ·

5. L. Desorgher, E. O. Flückiger, M. Gurtner, M. R. Moser, and R. Bütikofer, "Atmocosmics: a geant 4 code for computing the interaction of cosmic rays with the earth›s atmosphere," International Journal of Modern Physics A, vol. 20, no. 29, pp. 6802–6804, 2005. · ·

6. I. G. Usoskin and G. A. Kovaltsov, "Cosmic ray induced ionization in the atmosphere: full modeling and practical applications," Journal of Geophysical Research D, vol. 111, Article ID D21206, 2006. · ·

7. A. Mishev and P. I. Y. Velinov, "Atmosphere ionization due to cosmic ray protons estimated with Corsika code simulations," Comptes Rendus de L›Academie Bulgare des Sciences, vol. 60, no. 3, pp. 225–230, 2007.

8. P. I. Y. Velinov, A. Mishev, and L. Mateev, "Model for induced ionization by galactic cosmic rays in the Earth atmosphere and ionosphere," Advances in Space Research, vol. 44, no. 9, pp. 1002–1007, 2009. · ·

9. P. I. Y. Velinov, G. Nestorov, and L. Dorman, Cosmic Ray Influence on the Ionosphere and on the Radio-Wave Propagation, Bulgarian Academy of Sciences Publishing House, Sofia, Bulgaria, 1974.

10. H. S. Porter, C. H. Jackman, and A. E. S. Green, "Efficiencies for production of atomic nitrogen and oxygen by relativistic proton

impact in air," The Journal of Chemical Physics, vol. 65, no. 1, pp. 154–167, 1976.

11.   D. Heck, J. Knapp, J. N. Capdevielle, et al., "CORSIKA: a monte carlo code to simulate extensive air showers," Forschungszentrum Karlsruhe Report FZKA 6019, 1998.

12.   G. Battistoni, S. Muraro, P. R. Sala, et al., "The FLUKA code: description and benchmarking," inProceedings of theHadronic Shower Simulation Workshop, M. Albrow and R. Raja, Eds., vol. 896 of AIP Conference Proceeding, pp. 31–49, September 20062007.

13.   S. Ostapchenko, "QGSJET-II: towards reliable description of very high energy hadronic interactions,"Nuclear Physics B, vol. 151, no. 1, pp. 143–146, 2006. · ·

14.   R. Vainio, L. Desorgher, D. Heynderickx, et al., "Dynamics of the Earth›s particle radiation environment," Space Science Reviews, vol. 147, no. 3-4, pp. 187–231, 2009. · ·

15.   A. Mishev, P. I. Y. Velinov, and L. Mateev, "Atmospheric ionization due to solar cosmic rays from 20 January 2005 calculated with Monte Carlo simulations," Comptes Rendus de L›Academie Bulgare des Sciences, vol. 63, no. 11, pp. 1635–1642, 2010.

16.   I. G. Usoskin, G. A. Kovaltsov, I. A. Mironova, A. J. Tylka, and W. F. Dietrich, "Ionization effect of solar particle GLE events in low and middle atmosphere," Atmospheric Chemistry and Physics, vol. 11, no. 5, pp. 1979–1988, 2011. · ·

17.   A. L. Mishev, P. I. Y. Velinov, L. Mateev, and Y. Tassev, "Ionization effect of solar protons in the Earth atmosphere—case study of the 20 January 2005 SEP event," Advances in Space Research, vol. 48, no. 7, pp. 1232–1237, 2011.

18.   A. Mishev, P. I. Y. Velinov, and L. Mateev, "Ion production rate profiles in the atmosphere due to solar energetic particles on 28 october 2003 obtained with CORSIKA 6.52 simulations," Comptes Rendus de L›Academie Bulgare des Sciences, vol. 64, no. 6, pp. 859–866, 2011.

19.   R. Butikofer, E. O. Fluckiger, L. Desorgher, and M. R. Moser, "The extreme solar cosmic ray particle event on 20 January 2005 and its influence on the radiation dose rate at aircraft altitude," Science of the Total Environment, vol. 391, no. 2-3, pp. 177–183, 2008. · ·

20. N. K. Bostanjyan, A. A. Chilingarian, V. S. Eganov, and G. G. Karapetyan, "On the production of highest energy solar protons at 20 January 2005," Advances in Space Research, vol. 39, no. 9, pp. 1456–1459, 2007. · ·

21. C. Plainaki, A. Belov, E. Eroshenko, H. Mavromichalaki, and V. Yanke, "Modeling ground level enhancements: event of 20 January 2005," Journal of Geophysical Research A, vol. 112, no. 4, Article ID A04102, 2007. · ·

22. R. A. Mewaldt, M. D. Looper, C. M. S. Cohen, et al., "Solar-particle energy spectra during the large events of October-November 2003 and January 2005," in Proceedings of the 29th International Cosmic Ray Conference, vol. 1, pp. 111–114, Pune, India, 2005.

23. V. S. Makhmutov, G. A. Bazilevskaya, B. B. Grozdevsky, et al., "Solar cosmic ray spectra in the 20 January GLE: comparison of simulations with ballon and neutron monitor observations," in Proceedings of the 31th International Cosmic Ray Conference, pp. 1–4, Lodz, Poland, 2009.

24. A. Mishev and P. I. Y. Velinov, "Effects of atmospheric profile variations on yield ionization function Y in the atmosphere," Comptes Rendus de L›Academie Bulgare des Sciences, vol. 61, no. 5, pp. 639–644, 2008.

25. A. L. Mishev and P. Velinov, "The effect of model assumptions on computations of cosmic ray induced ionization in the atmosphere," Journal of Atmospheric and Solar-Terrestrial Physics, vol. 72, no. 5-6, pp. 476–481, 2010. · ·

26. M. Calisto, I. Usoskin, E. Rozanov, and T. Peter, "Influence of Galactic Cosmic Rays on atmospheric composition and dynamics," Atmospheric Chemistry and Physics, vol. 11, no. 9, pp. 4547–4556, 2011. · ·

27. L. I. Dorman, Cosmic Rays in the Earth›s Atmosphere and Underground, Kluwer Academic, Dordrecht, The Netherlands, 2004.

28. B. Funke, A. Baumgaertner, M. Calisto et al., "Composition changes after the «halloween» solar proton event: the High-Energy Particle Precipitation in the Atmosphere (HEPPA) model versus MIPAS data intercomparison study," Atmospheric Chemistry and Physics Discussions, vol. 11, no. 3, pp. 9407–9514, 2011. · ·

29. A. Krivolutsky, A. Kuminov, and T. Vyushkova, "Ionization of the atmosphere caused by solar protons and its influence on ozonosphere of the Earth during 1994–2003," Journal of Atmospheric and Solar-Terrestrial Physics, vol. 67, no. 1-2, pp. 105–117, 2005. · ·

30. A. A. Krivolutsky, A. V. Klyuchnikova, G. R. Zakharov, T. Y. Vyushkova, and A. A. Kuminov, "Dynamical response of the middle atmosphere to solar proton event of July 2000: three-dimensional model simulations," Advances in Space Research, vol. 37, no. 8, pp. 1602–1613, 2006. · ·

31. A. Damiani, M. Storini, M. Laurenza, and C. Rafanelli, "Solar particle effects on minor components of the Polar atmosphere," Annales Geophysicae, vol. 26, no. 2, pp. 361–370, 2008.

32. C. H. Jackman, D. R. Marsh, F. M. Vitt et al., "Short- and medium-term atmospheric constituent effects of very large solar proton events," Atmospheric Chemistry and Physics, vol. 8, no. 3, pp. 765–785, 2008.

33. C. H. Jackman, D. R. Marsh, F. M. Vitt et al., "Northern Hemisphere atmospheric influence of the solar proton events and ground level enhancement in January 2005," Atmospheric Chemistry and Physics, vol. 11, no. 13, pp. 6153–6166, 2011.

34. A. Mishev, P. I. Y. Velinov, and L. Mateev, "Atmospheric ionization due to solar cosmic rays from 20 January 2005 calculated with Monte Carlo simulations," Comptes Rendus de L›Academie Bulgare des Sciences, vol. 63, no. 11, pp. 1635–1642, 2010.

35. P. I. Y. Velinov, S. Asenovski, and L. Mateev, "Simulation of cosmic ray ionization profiles in the middle atmosphere and lower ionosphere on account of characteristic energy intervals," Comptes rendus de l'Academie bulgare des Sciences, vol. 64, no. 9, pp. 1303–1310, 2011.

36. K. Kudela, M. Storini, M. Y. Hofer, and A. Belov, "Cosmic rays in relation to space weather," Space Science Reviews, vol. 93, no. 1-2, pp. 153–174, 2000. · ·

37. A. L. Mishev and J. N. Stamenov, "Present status and further possibilities for space weather studies at BEO Moussala," Journal of Atmospheric and Solar-Terrestrial Physics, vol. 70, no. 2-4, pp. 680–685, 2008. · ·

38.   L. I. Miroshnichenko, "Solar cosmic rays in the system of solar-terrestrial relations," Journal of Atmospheric and Solar-Terrestrial Physics, vol. 70, no. 2-4, pp. 450–466, 2008.

39.   A. L. Mishev, "A study of atmospheric processes based on neutron monitor data and Cherenkov counter measurements at high mountain altitude," Journal of Atmospheric and Solar-Terrestrial Physics, vol. 72, no. 16, pp. 1195–1199, 2010

Citations

CHAPTER 1

Pöschl U (2012) Multi-stage open peer review: scientific evaluation integrating the strengths of traditional peer review with the virtues of transparency and self-regulation. *Front. Comput. Neurosci.* 6:33, doi: 10.3389/fncom.2012.00033.

CHAPTER 2

Alex Guenther, "Biological and Chemical Diversity of Biogenic Volatile Organic Emissions into the Atmosphere," ISRN Atmospheric Sciences, vol. 2013, Article ID 786290, 27 pages, 2013. doi:10.1155/2013/786290.

CHAPTER 3

Lelia N. Hawkins and Lynn M. Russell, "Polysaccharides, Proteins, and Phytoplankton Fragments: Four Chemically Distinct Types of Marine Primary Organic Aerosol Classified by Single Particle Spectromicroscopy," Advances in Meteorology, vol. 2010, Article ID 612132, 14 pages, 2010. doi:10.1155/2010/612132.

CHAPTER 4

Gimeno L (2013) Grand challenges in atmospheric science. Front. Earth Sci. 1:1. doi: 10.3389/feart.2013.00001.

CHAPTER 5

Lili Wang, Nan Zhang, Zirui Liu, Yang Sun, Dongsheng Ji, and Yuesi Wang, "The Influence of Climate Factors, Meteorological Conditions, and Boundary-Layer Structure on Severe Haze Pollution in the Beijing-Tianjin-Hebei Region during January 2013,"Advances in Meteorology, vol. 2014, Article ID 685971, 14 pages, 2014. doi:10.1155/2014/685971.

CHAPTER 6

Alexander Mishev, "Short- and Medium-Term Induced Ionization in the Earth Atmosphere by Galactic and Solar Cosmic Rays," International Journal of Atmospheric Sciences, vol. 2013, Article ID 184508, 9 pages, 2013. doi:10.1155/2013/184508.

Index